Signals and Communication Technology

This series is devoted to fundamentals and applications of modern methods of signal processing and cutting-edge communication technologies. The main topics are information and signal theory, acoustical signal processing, image processing and multimedia systems, mobile and wireless communications, and computer and communication networks. Volumes in the series address researchers in academia and industrial R&D departments. The series is application-oriented. The level of presentation of each individual volume, however, depends on the subject and can range from practical to scientific.

Indexing: All books in "Signals and Communication Technology" are indexed by Scopus and zbMATH

For general information about this book series, comments or suggestions, please contact Mary James at mary.james@springer.com or Ramesh Nath Premnath at ramesh.premnath@springer.com.

resolve occlusion issues and developed a masked face recognition model using Resnet-50 architecture. The model did well with the test, achieving a precision rate of 46% and an accuracy of 47%. The authors said that if obscured areas are cropped, the performance of the recognition can be further improved.

A hybrid model comprising a deep learning model, K-Nearest Neighbor algorithm, and local binary patterns was proposed by Vu et al. [24] for facial features extraction from the retinal region. The team obtained an F1 score 87 using their data set named COMASK20. The authors recommended that future research focuses on using lightweight, portable, and embedded models. A hybrid model of CNN, Bag of Features, and multi-layer perceptron through the transfer learning technique was presented in [25]. The researcher achieved 91% accuracy using the VGG16 architecture.

Another approach was presented in [11] by improving the residual network, which they named the RggNet. RggNet was proved effective on a small dataset of 65 identities selected from the VGGFACE2 dataset, with an accuracy of 60.8%, a precision of 77.70%, a recall rate of 51.9%, and an AUC value of 95.3. A different study published by Golwalkar and Mehendale [26] involves proposing a CNN known as FaceMaskNet, which the authors trained from scratch with their dataset of 2000 images of 13 people combined with Labelled Faces in the Wild (LFW) set. The authors' accuracy rate in identifying faces with masks was 88.9%.

Eman et al. [27] introduced a new approach to masked face recognition. They incorporated an improved principal component analysis for feature extraction with a deep learning algorithm for detecting already trained landmarks. Finally, they utilized face detection using the K-Nearest Neighbor algorithm. They did so while simultaneously achieving a remarkable 97% recognition accuracy rate, but only implicitly. The comparative study of the effectiveness of three pre-trained CNN architectures, Resnet-50, AlexNet, and Inception V3, was done independently by Esi Nyarko et al. in 2022. The AlexNet model gained a success rate of 95. The InceptionV3 model followed 95.5% accuracy for human-level performance, while the Resnet 50 model achieved 97. The authors found that using their dataset, this model achieves roughly 5% accuracy.

3 Research Methodology

3.1 Research Design

The study used an experimental method to establish the hybrid masked face recognition models. VGG16 and MobileNetV2 networks were adopted for this research, and the proposed architecture is depicted in Fig. 3. The development of the proposed model was initiated by tuning the hyperparameters of chosen CNN architectures. The tuning was achieved by "freezing" the layers containing generically trained image characteristics and then optimizing the hyperparameters of the later layers to allow them to learn the characteristics of masked faces. The fully connected layers were replaced with randomly initialized ones. All the convolutional layers employed the rectified linear unit (ReLU) as the activation function because of its computational fastness and ability to handle the vanishing gradient problem [28].

Three hyperparameters were tuned: the learning rate, the optimizer, and the dropout rate. To select the optimal hyperparameters from the set, the study adopted the grid search technique to choose the parameter values based on the validation accuracy. This process entailed going through a range of learning rates, various optimizers, and the range of the dropout rates that would improve the performance of the models. This paper explored two optimization options: Adam and Stochastic Gradient Descent (SGD). Other experimental features were dropout rates of 0.2 and 0.5, and learner rates equal 0.0001 and 0.001. All models were trained to 100 epochs; a decision was made to choose a middle ground where the model converges well but at the same time does not overfit. Also, the sample inputs were loaded to the models in batches of 64.

The fine-tuned models were then integrated with the linear multiclass SVM to develop two hybrid models: VGG-16 + LMSVM and MobileNetV2 + LMSVM. During integration phase, the CNN models that were fine-tuned were used to extract feature embedding of the training, validation, and test input images. These extracted features were then flattened to vectors and input to the linear multiclass support vector machine classifier where the recognition task was performed. By leveraging the discriminative features learned by the CNN models, the multiclass SVM classifier became adept at accurately classifying masked facial images into their respective classes. The linear SVM classifier was trained using the one versus rest classifier and a C parameter 100 (Fig. 1).

3.2 Dataset

The masked VGG2 face dataset was used to train, validate, and test each model. As seen in Fig. 2, every image in this dataset has a face mask covering it, and some of the photos also have head hats and eyeglasses on. Blending the covered part with the rest of the face is the main challenge in recognizing faces with covered parts. To address this, the images were first cropped at a height of 0.9L by the findings of [29]. After that, they were divided into groups to train, evaluate, and replicate the models. Fifteen percent of the images were set for the test and 15% for the validation, while 70% were provided for training (Fig. 2).

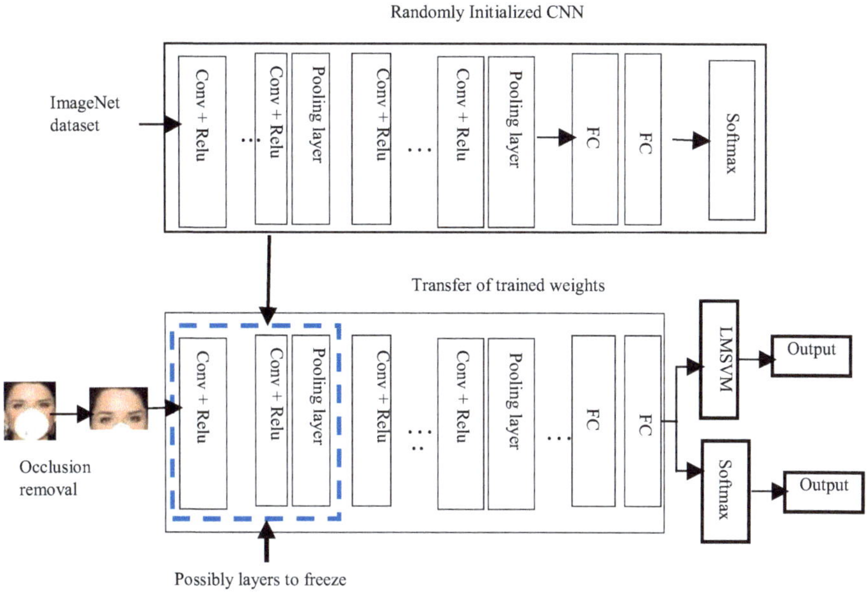

Fig. 1 Proposed architecture

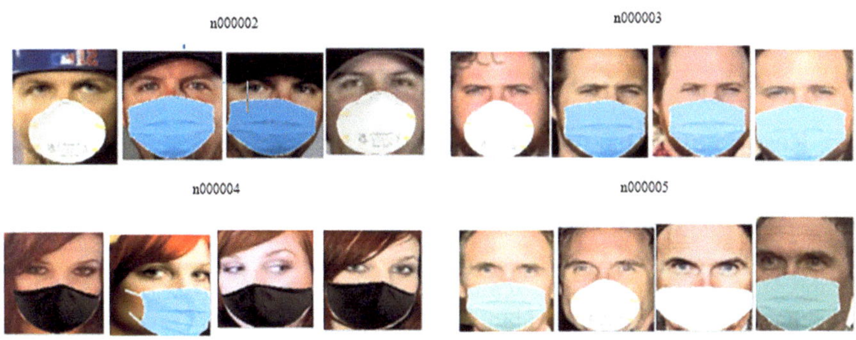

Fig. 2 Sample dataset

3.3 *Performance Evaluation*

Recognition accuracy, precision, recall, and F1 score were used as criteria for assessing the developed models. The recognition accuracy was computed by summing up the number of accurately predicted images and dividing the result by the total number of images given to the model for testing, as expressed in Eq. (2).

$$\text{Accuracy} = \frac{\text{TP} + \text{TN}}{\text{TP} + \text{TN} + \text{FP} + \text{FN}} \tag{2}$$

Precision, on the other hand, helped assess the percentage of the hybrid model that was accurate out of all facial images it had recognized. It is calculated by dividing the actual positive values by the sum of true positive and false positive values, as illustrated in Eq. (3).

$$\text{Precision} = \frac{\text{TP}}{\text{TP} + \text{FP}} \tag{3}$$

Recall is another measure that can describe the correctness of the positive predictions made by the hybrid model regarding all potential positive predictions. This measure was calculated by dividing the actual positive values by the total of actual positive and false negative values, as shown in Eq. (4).

$$\text{Recall} = \frac{\text{TP}}{\text{TP} + \text{FN}} \tag{4}$$

The F1 score harmoniously combines precision and recall to assess the effectiveness of the model in accurately identifying positive instances. It is achieved by computing the weighted average of the recall and precision values of the model as shown in Eq. (5).

$$\text{F1 Score} = \frac{2 \times \text{Precision} \times \text{Recall}}{\text{Precision} + \text{Recall}} \tag{5}$$

4 Results and Discussion

4.1 Hyperparameter Tuning

In this study, a comprehensive optimization approach was used. The researchers experimented with different combinations of hyperparameters for every model, covering all the bases. The validation accuracy evaluation metric was employed to appraise the efficacy of these combinations. The validation accuracy attained by every combination of the three hyperparameter values for all models is shown in Table 1, which is shown below. All the models performed their best with the configuration, including the Adam optimizer, with a learning rate 0.0001 and a dropout rate of 0.5.

Table 1 A comparative table showing the validation accuracies of each hyperparameter value combination

Pre-trained architecture	Optimizer	Dropout rate	Learning rate	Validation accuracy (%)
VGG16	Adam	0.2	0.001	85.30
		0.2	0.0001	92.76
		0.5	0.001	47.96
		0.5	0.0001	95.82
	SGD	0.2	0.001	94.30
		0.2	0.0001	86.31
		0.5	0.001	91.34
		0.5	0.0001	71.83
MobileNetV2	Adam	0.2	0.001	87.54
		0.2	0.0001	89.37
		0.5	0.001	87.29
		0.5	0.0001	91.24
	SGD	0.2	0.001	87.07
		0.2	0.0001	68.85
		0.5	0.001	84.46
		0.5	0.0001	58.02

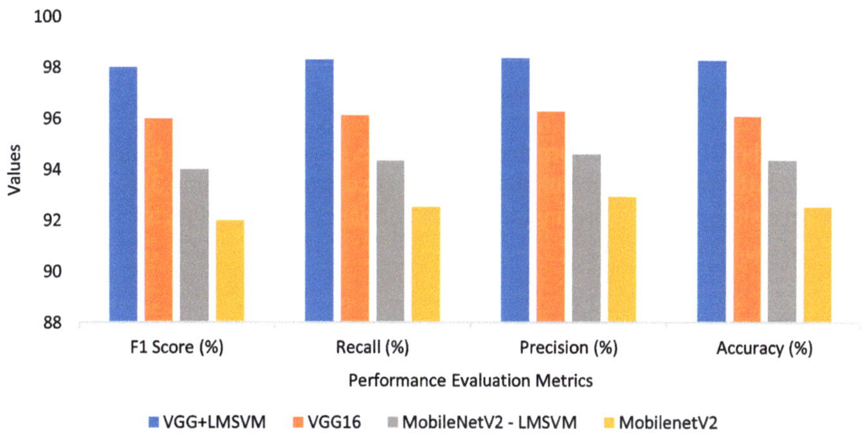

Fig. 3 A clustered column chart comparing the performance of the hybrid models to their standalone variants

4.2 Performance Evaluation of the Developed Models

Four evaluation measures were employed to evaluate the performance of the developed models: recall, precision, F1 score, and accuracy. The clustered bar column chart illustrated in Fig. 3 provides a visual manner in which the metrics values of the four developed models can be compared. The results show that the hybrid models are superior to the standalone CNN models across all the measures for all the cases. This performance is attributed to the robustness of SVM against overfitting and the kernel trick. The results showed that by integrating the LMSVM into the VGG16 model, precision improved by about 2.18%. Likewise, the performance of the hybrid model MobileNetV2 + LMSVM was slightly higher than MobileNetV2, with a 1.82% improvement in precision compared to the MobileNetV2 network. As noted, these features show just how well the hybrid models excel in finding the positive instances while simultaneously reducing the number of false positives with similar levels of precision.

The hybrid model VGG16 + LMSVM showed a boost of 2.17% in terms of recall measures. Likewise, the recall of the hybrid model MobileNetV2 + LMSVM was better than that of the MobileNetV2 network by around 1.82%. These results show to what extent hybrid models are suited to detect concealed faces regardless of the identity class properly and, at the same time, minimize false negatives for each of them.

The recognition accuracy metric value for each model was determined, and the overall tally was made in the form of a table. From the results shown in Table 2, the combined models with LMSVM (VGG16 + LMSVM and MobileNetV2 + LMSVM) yield better results than the pure CNN model (VGG16 and MobileNetV2) alone. From this, it can be deduced that high performance in recognizing the masked face images used during the testing phase is realizable when the LMSVM classifiers are fused with the CNN models. The improvement of test accuracy for the hybrid models means that with the given set of sample images, it is possible to obtain better predictions from the SVM classifiers with the help of effective use of features from the CNNs.

The hybrid models' performance against the standalone CNN created for this study, and the previously examined methods is contrasted in Table 3. Recall, accuracy, precision, and F1 score are the four performance evaluation metrics used in the comparison.

5 Conclusion and Future Work

The conducted experiments validated the results found in the literature review, demonstrating that the fine-tuned CNN performs better when combined with a linear multiclass support vector machine than when used alone. About 98.27% recognition accuracy, 98.30% recall, 98.37% precision, and 98% F1 score were attained by the hybrid VGG16 + LMSVM model. Similarly, the hybrid MobileNetV2 + LMSVM

Table 2 A recognition accuracy comparative table

Model	Recognition accuracy (%)
VGG16 + LMSVM	98.27
VGG16	96.09
MobileNetV2 + LMSVM	94.36
MobileNetV2	92.54

Table 3 A performance comparative table

Models	Accuracy (%)	Precision (%)	Recall (%)	F1 score (%)
VGG16 + LMSVM	98.27	98.37	98.30	98
MobileNetV2 + LMSVM	94.36	94.58	94.35	94
VGG16	96.09	96.28	96.13	96
MobileNetV2	92.54	92.93	92.53	92
ResNet50 [23]	47	46.13	–	–
DNN [22]	68.26			
CNN + KNN + MLP [24]	–	–	–	87
CNN + BOG + MLP [25]	91.3	–	–	–
Modified Residual Network architecture [11]	60.8	77.70		
CNN [26]	88.9	–	–	–
Pre-trained Landmark detection deep learning algorithm + PCA + KNN [27]	97	–	–	–
Resnet-50 [30] AlexNet [30] InceptionV3 [30]	97.5 95.5 95.7			

model yielded a 94.36% recognition accuracy, 94.35% recall, 94.58% precision, and 94% F1 score. In contrast, 96.09% recognition accuracy, 96.13% recall, 96.28% precision, and 97% F1 score were attained by the VGG16 model. About 92.54% recognition accuracy, 92.93% recall, 92.53% precision, and 93% F1 score were attained by the MobileNetV2 model.

These findings show that, in comparison to using standalone CNN models, hybridizing optimized CNN models with LMSVM improves performance and accuracy in identifying masked faces. Therefore, the hybrid model can be deployed in healthcare sectors to seamlessly recognize patients and staff wearing surgical masks without compromising safety guidelines. In addition, it can be incorporated into access control systems to allow workers to be authenticated even when wearing face masks for various reasons, including safety against infectious diseases.

The performance impact of combining another pre-trained convolutional neural network with a linear multiclass support vector machine classifier in masked face recognition can be investigated further in the future. In addition, more studies can be done to enhance the precision and accuracy of face recognition under different occlusion conditions.

References

1. Shepley, A.J.: Deep learning for face recognition: a critical analysis. ArXiv Prepr., ArXiv190712739 (2019)
2. Bykov, A., Voronov, V., Voronova, L., Zharov, I.: Web application development for biometric identification system based on neural network face recognition. In: 2020 Systems of Signals Generating and Processing in the Field of on Board Communications, pp. 1–6. IEEE (2020)

3. Hansstein, F.V., Echegaray, F.: Exploring motivations behind pollution-mask use in a sample of young adults in urban China. Glob. Health. **14**, 1–10 (2018)
4. Desai, A.N., Mehrotra, P.: Medical masks. JAMA. **323**, 1517–1518 (2020)
5. Jaiswal, S.D., Verma, A.K., Mukherjee, A.: Mask-up: investigating biases in face re-identification for masked faces. ArXiv Prepr., ArXiv240213771 (2024)
6. Almabdy, S., Elrefaei, L.: Deep convolutional neural network-based approaches for face recognition. Appl. Sci. **9**, 4397 (2019)
7. Bhatt, D., Patel, C., Talsania, H., Patel, J., Vaghela, R., Pandya, S., Modi, K., Ghayvat, H.: CNN variants for computer vision: history, architecture, application, challenges and future scope. Electronics. **10**, 2470 (2021)
8. Deepak, S., Ameer, P.: Automated categorization of brain tumor from MRI using CNN features and SVM. J. Ambient Intell. Humaniz. Comput. **12**, 8357–8369 (2021)
9. Awais, M., Chiari, L., Ihlen, E.A., Helbostad, J.L., Palmerini, L.: Classical machine learning versus deep learning for the older adults free-living activity classification. Sensors. **21**, 4669 (2021)
10. O'Mahony, N., Campbell, S., Carvalho, A., Harapanahalli, S., Hernandez, G.V., Krpalkova, L., Riordan, D., Walsh, J.: Deep learning vs. traditional computer vision. In: Advances in Computer Vision: Proceedings of the 2019 Computer Vision Conference (CVC), vol. 1, pp. 128–144. Springer (2020)
11. Kumar, M., Mann, R.: Masked face recognition using deep learning model. In: 2021 3rd International Conference on Advances in Computing, Communication Control and Networking (ICAC3N), pp. 428–432. IEEE (2021)
12. Setiowati, S., Franita, E.L., Ardiyanto, I., et al.: A review of optimization method in face recognition: comparison deep learning and non-deep learning methods. In: 2017 9th International Conference on Information Technology and Electrical Engineering (ICITEE), pp. 1–6. IEEE (2017)
13. Alzubaidi, L., Zhang, J., Humaidi, A.J., Al-Dujaili, A., Duan, Y., Al-Shamma, O., Santamaría, J., Fadhel, M.A., Al-Amidie, M., Farhan, L.: Review of deep learning: concepts, CNN architectures, challenges, applications, future directions. J. Big Data. **8**, 1–74 (2021)
14. Nogueira, K., Penatti, O.A., Dos Santos, J.A.: Towards better exploiting convolutional neural networks for remote sensing scene classification. Pattern Recognit. **61**, 539–556 (2017)
15. Wang, M., Deng, W.: Deep face recognition: a survey. Neurocomputing. **429**, 215–244 (2021)
16. Kaya, Y., Gürsoy, E.: A MobileNet-based CNN model with a novel fine-tuning mechanism for COVID-19 infection detection. Soft Comput. **27**, 5521 (2023)
17. Patel, R., Chaware, A.: Comparative analysis of hyperparameter tuned convolutional neural networks for classification of diabetic retinopathy. Int. J. Intell. Syst. Appl. Eng. **11**, 2147–6799 (2023)
18. Nwankpa, C., Ijomah, W., Gachagan, A., Marshall, S.: Activation functions: comparison of trends in practice and research for deep learning. ArXiv Prepr., ArXiv181103378 (2018)
19. Cervantes, J., Garcia-Lamont, F., Rodríguez-Mazahua, L., Lopez, A.: A comprehensive survey on support vector machine classification: applications, challenges and trends. Neurocomputing. **408**, 189–215 (2020)
20. Kumar, B., Sobti, P., Vibhandik, O.S., Kumar, A., Sharma, V., Vyas, O.: An architecture comprising of convolutional neural network (CNN)-support vector machine (SVM) and CNN-SoftMax for cricket shots classification. In: 2022 IEEE 19th India Council International Conference (INDICON), pp. 1–6. IEEE (2022)
21. Liu, S., Tang, X., Wang, D.: Facial expression recognition based on Sobel operator and improved CNN-SVM. In: 2020 IEEE 3rd International Conference on Information Communication and Signal Processing (ICICSP), pp. 236–240. IEEE (2020)
22. Qiu, H., Gong, D., Li, Z., Liu, W., Tao, D.: End2end occluded face recognition by masking corrupted features. IEEE Trans. Pattern Anal. Mach. Intell. **44**, 6939–6952 (2021)
23. Mandal, B., Okeukwu, A., Theis, Y.: Masked face recognition using resnet-50. ArXiv Prepr., ArXiv210408997 (2021)

24. Vu, H.N., Nguyen, M.H., Pham, C.: Masked face recognition with convolutional neural networks and local binary patterns. Appl. Intell. **52**, 5497–5512 (2022)
25. Hariri, W.: Efficient masked face recognition method during the covid-19 pandemic. Signal Image Video Process. **16**, 605–612 (2022)
26. Golwalkar, R., Mehendale, N.: Masked-face recognition using deep metric learning and FaceMaskNet-21. Appl. Intell. **52**, 13268–13279 (2022)
27. Eman, M., Mahmoud, T.M., Ibrahim, M.M., Abd El-Hafeez, T.: Innovative hybrid approach for masked face recognition using pretrained mask detection and segmentation, robust PCA, and KNN classifier. Sensors. **23**, 6727 (2023)
28. Mogan, J.N., Lee, C.P., Lim, K.M., Muthu, K.S.: VGG16-MLP: gait recognition with fine-tuned VGG-16 and multilayer perceptron. Appl. Sci. **12**, 7639 (2022)
29. Li, Y., Guo, K., Lu, Y., Liu, L.: Cropping and attention based approach for masked face recognition. Appl. Intell. **51**, 3012–3025 (2021)
30. Nyarko, B.N.E., Bin, W., Zhou, J., Agordzo, G.K., Odoom, J., Koukoyi, E.: Comparative analysis of AlexNet, Resnet-50, and Inception-V3 models on masked face recognition. In: 2022 IEEE World AI IoT Congress (AIIoT), pp. 337–343. IEEE (2022)

The Evolution in Process Automation and Control Using Artificial Intelligence (AI) and Machine Learning (ML) in Fourth Industrial Revolution (4IR)

Puramanathan Naidoo

1 Introduction

The field of process automation and control has transformed substantially over the past decades. Simple mechanical systems that were manually monitored and controlled have been transformed into advanced digital solutions, with the introduction of a Human Machine Interface (HMI). The HMI makes provision for operations personnel to interact with the process remotely, where plant conditions are displayed, allowing the operator to manually or automatically control the process.

Process automation and control refers to the detection of a physical process by the measuring instrument, which provides a transmission signal. This signal is applied as an input to the automation or control system, where it is processed in an algorithm to generate an output signal. This transmission signal is applied to the Final Correcting Element (FCE). Automation normally refers to a repetitive process comprising mainly binary signals, where limited or minimal analogue signals are processed, e.g. a conveyor system. On the other hand, control normally refers to a dynamic process comprising mainly analogue signals, where limited or minimal digital signals are processed, e.g. a tank level system with a control valve. The technology is used to eliminate manual effort in the process under normal operating conditions. Manual intervention is only required in the event of unusual circumstances, e.g. when a tank overflows, for immediate reaction, to settle the process and minimise any effects on the process.

The introduction of the Fourth Industrial Revolution (4IR) gave rise to Artificial Intelligence (AI) and Machine Learning (ML) where process automation has taken a substantial jump forward, allowing for more innovative measuring and control

P. Naidoo (✉)
Mangosuthu University of Technology, Durban, South Africa
e-mail: pnaidoo@mut.ac.za

P. Naidoo, M. Sibanda (eds.), *Measurement, Automation, and Control in Artificial Intelligence and Machine Learning*, Signals and Communication Technology, https://doi.org/10.1007/978-3-031-91802-5_4

methods, to further optimise processes. AI makes provision for a system to simulate human behaviour whilst ML makes provision the system to emulate a response from archived data, to further optimise processes. ML is a subset of AI, where configured algorithms allow systems to generate signals from stored data and perform decision-making processes similarly to human reasoning. This chapter aims to provide an overview of this evolution [1].

2 Considerations to Implement 4IR, AI and ML

This section provides some overview in international and national level considerations, over the past years, to gauge actual implementation. As reported in 2016 by the World Economic Forum, a distinct paradigm shift was present across all works of life. In general, revised business approaches were introduced, to optimise operations in production, including the logistics of transportation. In terms of society, we have transformed the way we work and interact. At national level, governments have made positive changes to the sectors of education, health, and social security, among others. Many manual processes were replaced with automated systems, creating a more user-friendly and convenient atmosphere for the citizens, including improved environment control. Previously, environmental aspects were secondary but later found that deteriorating environmental conditions ultimately has bearing on the health of the citizens. In the presence of delayed development in some cases, the adoption of 4IR, AI and ML, technologies creates uncertainty in how this evolution must be driven, with the participation of all the relevant stakeholders, especially since the pace at which this transformation takes place varies amongst academia/ research, business, government, and society. All stakeholders need to interact and implement changes at the same level, which is very often based on knowledge and skills, to understand the evolving trends. A low level of leadership was identified, and there is a lack of understanding across all sectors, requiring a revised way of economics, society, and political structures, to adapt to 4IR. It was found both at the national and global levels that there is a lack of structured framework to manage the application of this evolution and manage the interruption that may be caused. It was further found that there is a global lack of consistency, confidence, and common ground to recognise the opportunities and challenges of 4IR, thereby infringing on the processes to build capacity amongst individuals and communities and prevent any repercussion that may be forthcoming [2].

As reported in 2020 in line with local government preparing for 4IR in South Africa, it was found that 4IR will primarily affect jobs and skills requirements. "Five years from now, over one-third of skills (35%) that are considered important in today's workforce will have changed". At the same time, the knowledge gained in a classroom will not be required by the time the graduate enters the workplace. It was found that the most critical skills needed in 2020 will require qualities like problem-solving techniques, critical thinking, and emotional intelligence, amongst many others, to adapt to this evolution. Research has shown that AI and ML will contribute to unemployment, at the same time create a demand for new jobs and skill sets, thereby upgrading workers to higher levels instead of becoming unemployed. Nicholas Davis, an originator of the phrase, the "Fourth Industrial Revolution" stated that "If we talk about it and work on it together at the beginning of the revolution, maybe we will avoid the mistakes made in previous revolutions, which resulted in massive exclusion and human rights violations, environmental damage and regulatory lag". Commencing at government level, there is a need for a proactive approach instead of a reactive situation when the technologies are ready to be used. "Rapidly improving access to electricity should be a key policy priority. Governments should view energy security as a function of investment in renewables and the foundation for future growth" [3, 4].

3 The Process Automation and Control System

A process automation and control system comprises four main components, as shown in Fig. 1 below. Commencing at the process, this refers to the physical plant, where a specific variable needs to be measured, e.g. level, pressure, and flow. Measurement of the process may be required for merely knowing the status of the variable, in which case there is no need for any automation of control, but only an indicator. Where the process needs to be measured and controlled, this section discusses related aspects. The physical status of the plant is detected by a sensor in the measuring instrument, converting it to a transmission signal, referred to as the process variable. The process variable is generated as an input to the control instrument. The control instrument consists of a control or automation algorithm that defines a set point, to maintain a desired process condition. Depending on the error signal between the process variable and a set point, an output signal is generated by the control instrument, referred to as the manipulated variable. The manipulated variable is generated as an input to the final correcting element, which could be a control valve or a variable speed drive, where the physical process is regulated. The

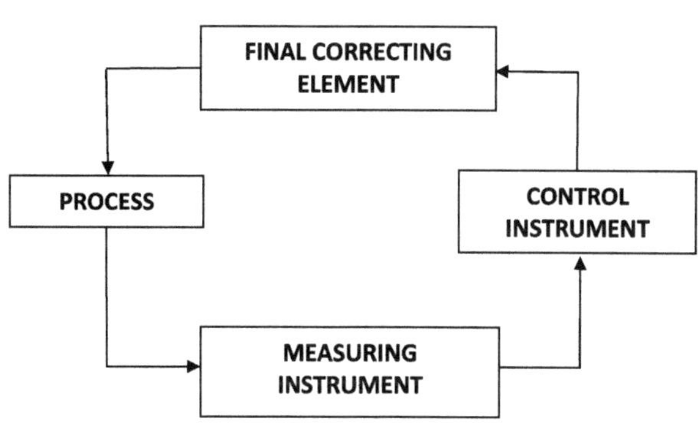

Fig. 1 Components of a process automation and control system [5, 6]

regulation of the process medium aims to maintain a zero-error signal between the process variable and the set point, to maintain the desired process condition as per the set point.

3.1 Measuring Instrument

Process measuring instruments functioning on AI technology have substantially augmented the capabilities into the control or automation system. Through AI, the instruments do not only transmit real-time data, representative of the physical process condition, to the control instrument, but they can evaluate the measured data, thereby creating the ability to predict plant behaviour and detect abnormal plant conditions. This is normally accomplished in the control or automation system. The early detection of abnormal plant behaviour can prevent equipment failure, which contributes to effective predictive maintenance, thereby improving machine repeatability. AI algorithms can automatically adjust parameters of the measuring instrument to decrease deadtime, thereby optimising the measurement of variables like temperature, pH, and conductivity, amongst others, where there is a delay in the output signal.

3.2 Automation/Control System

Traditionally, specific algorithms, within set limits, control the process variables to maintain desired process conditions. In this type of algorithm, the control architecture is confined to limitations, which may be mathematically based allowing concepts like 0.25 wave damping to determine how the final correcting element regulates the final process. Like in a Proportional, Integral, Derivative (PID) controller, the controller maintains the manipulated variable if the process variable is in line with set point or the desired value, thereby maintaining a steady input and output of the controller. If the physical process varies due to varying plant conditions or any related input disturbances, the process variable varies, to maintain desired plant conditions. As seen in Fig. 2 below, when the controller is switched from manual to automatic, the process variable, indicated by PV is automatically adjusted to be in line with the set point, indicated by SP. The response for the manipulated variable, indicated by MV is not represented to focus discussion on PV. The response of PV against SP indicates that the overshoot or undershoot, after the first, decreases with time. Using a data logger, the amplitude of the following overshoot or undershoot will be 0.25 of the previous one, hence described as 0.25 wave damping. It must also be noted that in such a control algorithm, the PID settings that produce a response as indicated do not necessarily mean that the same response will be achieved for other set point values. Furthermore, due to the mathematical

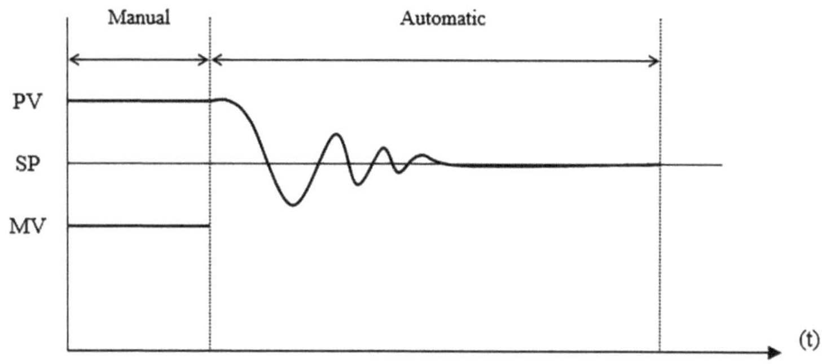

Fig. 2 Concept of 0.25 wave damping in a PID controller

reasoning, these oscillations cannot be avoided, for the controller to be tuned, using various methods. Hence the control strategy has some shortfalls as discussed.

However, an AI-based controller is knowledge-based, facilitating an algorithm to respond to different sets of rules, for different process conditions. With the introduction of AI, these systems function based on human reasoning or respond by learning from archived data, by continuously adapting in real time with process dynamics. One of the AI-based architectures is the Fuzzy Logic Controller (FLC). In this architecture, the two types of human knowledge are specified:

- Plant knowledge: Fuzzy IF-THEN rules describe the behaviour of the plant, e.g. "IF the pressure control valve opens, THEN the pressure of the process medium will increase", where "open" and "increase" are characterised by fuzzy sets.
- Control knowledge: Fuzzy control rules that state in which situations what control actions should be taken, e.g. "IF the process pressure decreases, THEN the output of the FLC should increase", where "decrease" and "increase" are characterised by fuzzy sets.

The structure of the adaptive FLC is classified into the following three categories:

- Indirect adaptive fuzzy control where the algorithm comprises a number of fuzzy systems constructed (initially) from the plant knowledge.
- Direct adaptive fuzzy control where the algorithm is a single fuzzy system constructed (initially) from the control knowledge.
- Combined indirect/direct adaptive fuzzy logic is a weighted average of the indirect and direct adaptive FC (therefore, both plant and control knowledge are used) [6, 7].

The overview of an AI-based control architecture is represented in Fig. 3 below, specifically the FLC. The various stages are explained briefly below.

- Stage 1: field sensors measure the process variables
- Stage 2: sensors transmit input data signals
- Stage 3: input data signals are assigned to membership functions

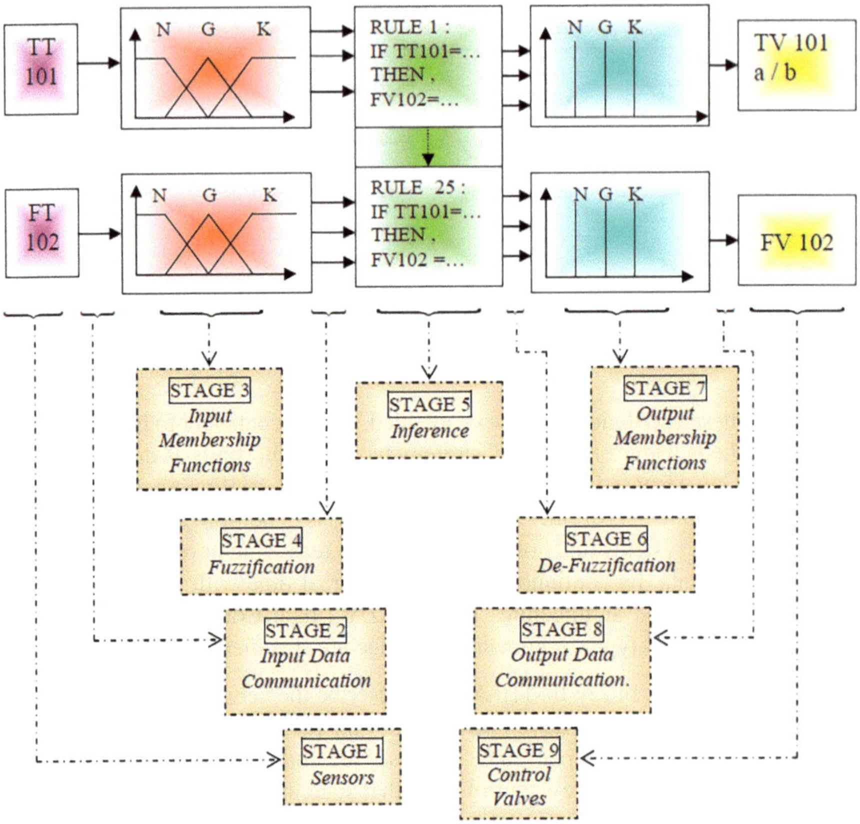

Fig. 3 Structure of a fuzzy logic controller [6, 7]

- Stage 4: membership functions are fuzzified, to adapt to "Intelligent interpretation"
- Stage 5: fuzzified inputs are applied to IF-THEN rules, generating fuzzified outputs
- Stage 6: fuzzified outputs are de-fuzzified, to adapt to other peripheral devices
- Stage 7: de-fuzzified values are assigned to membership functions
- Stage 8: output data transmission signals are generated to correcting devices
- Stage 9: correcting devices adjust process conditions to desired value

3.3 Final Correcting Element

The final correcting element can be in the form of Variable Speed Drives (VSD) or control valves, the latter being more common. In almost all process applications, a motor is driving a pump on a tank, or driving a gearbox on a conveyor or agitator, amongst others. VSDs can control a process in the absence of a controller, due to its ability to respond to feedback from an encoder fitted onto the motor. However, such installations may have limitations. Ideally, a dedicated process control system is required. Hence pneumatic control valves are commonly used in the process automation and control industry. Additional benefits are the fail-safe open and fail-safe close interlocks, in the case of emergencies, where a valve can return to a fully opened or fully closed position when required. Furthermore, pneumatic instrumentation is here to stay, in the age of 4IR, AI and ML. However, with valves, it is adapting and integrating with AI technology that will revolutionise the industry. Conventional control valves depend on a controller responding to a desired value or set point, to generate an out signal to the control valve, to regulate the process in accordance with process conditions. Control valves that are AI compliant can access real-time data from measuring instruments to determine the valve position, similarly like the way VSDs use the feedback signal from motor encoders to control the motor speed. In this way, the concept of repeatability is optimised in instruments of the control loop. ML algorithms can be used to predict process changes. AI-compliant valves have the added benefit to diagnose any anomalies in the valve. Although the control valve is one of the four main components in a control loop, it is purely a mechanical device. There are two factors that determine the position of the control valve, viz. the air signal and the spring tension. There are two types of valves, viz. Air-to-Open (ATO) and Air-to-Close (ATC). In an ATC valve, the spring tension keeps it opened, and the air signal closes it, between a 0% to 100% valve travel stroke. Due to the mechanical structure of the valve, that is functioning with instruments functioning on 4IR, AI or ML, it becomes critical to identify any mechanical failure, e.g. metal fatigue, cavitation, blockage, or leakage, to avoid any infringing of the control strategy. In this way, the control strategy and predictive maintenance procedures are optimised in the interest of overall business performance.

4 The Evolution of Process Automation and Control

The industrial revolution was a period of transforming economics, technology, and society as a whole. It marks the paradigm shift from manual labour-intensive systems to machine-driven systems. As factories grew, the manufacturing industry grew exponentially. This revolution created more jobs and people migrated from rural to urban areas in search of semi-skilled and skilled jobs, thereby shaping the economy. The industrial revolution, in the process, created other subsequent modernisation changes, to reach the stage of digitisation. There is significant change in the process automation and control landscape, including measurement, over the past decades. Previously, measurement was a physical determination to monitor or control a process, and measuring the instruments were introduced. These instruments varied from instruments that are for purely monitoring, e.g. pressure gauge, to instruments with transmission signals to be used for control, e.g. flow transmitter, to non-contact type instruments that do not make physical contact with the process to take measurements, e.g. ultrasonic transmitters, to the common contact type instruments that are installed in the process to take measurements. The following discussion provides overview of the four industrial revolutions.

4.1 First Industrial Revolution (1760–1840)

In this first phase of the industrial revolution in the eighteenth century, production of goods and products took place through semi-mechanical processes and manual labour was slowly being replaced by machines. This increased production targets, with greater efficiency. Automation at this stage was purely through mechanical systems, e.g. the steam engine, manually driven machines. The human element was always integral in all systems.

4.2 Second Industrial Revolution (1870–1914)

In this second phase of the industrial revolution, in the late nineteenth and early twentieth centuries, the production of electricity became the new source of power, thereby replacing many manual processes. Electricity revolutionised the process automation and control environment by facilitating the introduction of control systems. Machines that were manually operated were driven by electrical motors, resulting in increased production and reduced fatigue amongst the work force. Basic control devices like switches, timers, and counters improved automation and control processes.

4.3 Third Industrial Revolution (1950s–2000s)

In this third phase of the industrial revolution, in the mid-twentieth century, the process automation and control landscape were digitised. Computers were introduced followed by the Programmable Logic Controllers (PLC) in the 1960s. PLCs became the revolutionary change in process automation and control. Complex processes could be controlled programming the PLC through a computer. The program software resided in the computer, which was interfaced to the PLC to write a program for a process application and download it to the PLC. When the PLC was interfaced with the process, it became possible to control the plant in the absence of the human element. There was further development into Supervisory Control and Data Acquisition (SCADA), Distributed Control System (DCS), and MES (Manufacturing Execution Systems). DCS, SCADA, and MES systems made it possible for an operator to interface with the technology by means of an HMI and made possible the archiving of data for reference purposes at any time. This paradigm shift was followed by systems like Computer-Aided Design (CAD) and Computer-Aided Manufacturing (CAM), which made it possible to convert a drawing into a manufacturing process. The power of this revolution made it possible for a computer, with the relevant software programs, to execute any manufacturing process in real-time. Industrial robots were introduced in the 1970s to revolutionise the automotive and packaging industries, amongst many other rapid prototyping processes.

4.4 The Fourth Industrial Revolution: Intelligent Automation, AI, and ML

In this current fourth phase, that we refer to as 4IR, we witness advanced technologies like AI, ML, and IIoT. It represents a shift in how we live, work, and interact with each other and the world around us. AI and ML introduce intelligent systems for measurement, automation, and control, facilitating machines to extract data, learn from the data, and make intelligent decisions with minimum or no human involvement. However, all systems have a manual facility where the human element is required under extreme conditions. IIoT devices interface all systems in a business environment. Although process automation and control exist in the manufacturing process, IIoT takes it to another level where the finances, production, human resources, amongst others, all function as a fully integrated system. In this way, final statistics in a business process is more easily generated as compared to a segregated system, with the different departments generating their own statistics, to eventually consolidate all. 4IR is a new world of living and working.

5 The Fourth Industrial Revolution and the Emergence of AI and ML

4IR represents the phase that we are in, where all systems converge, in a manufacturing plant, introducing AI and ML. AI algorithms extract designated amounts of data in real time, analyse them, and transmit signals to machines for decision-making and to determine the running status of the machines. ML continuously learns from generated data, to determine patterns in how the machines function optimally. This revolution is transforming the manufacturing industry, amongst others, by integrating systems and making it possible for measuring instruments and final correcting elements to function in the absence of a controller. With AI and ML becoming an integral component in industrial systems, businesses can respond appropriately to market changes, improve product quality, increase production rates and most of all, and optimise customer service.

In the simplest term, day-to-day banking has optimised the way we interact with the bank. There is no need for cash transactions. Waiting in queues at the banks is a practice we no longer are at mercy of. Our cell phone has become that interface between the bank and us, like the way it has become an interface with almost every sphere of life, from knowing what the weather is to attending meeting online. We can plan our daily lives with the various apps on our cell phone. In the same way, a process plant in one country can be controlled from a cell phone from anywhere in the world.

5.1 Evolution of Industry 4.0

Prior to 4IR, IT provided support and infrastructure within the business units and primarily the administrative departments in industry, e.g. finance, human resources, logistics, amongst others. IT served the functions of data storage and network security and maintains administrative software. These IT systems functioned in isolation from process environment, where the engineering systems of the factory focused on the manufacturing, to primarily convert some raw material to the final product. On another level, IT systems coordinated production operations, again in isolation. IT structure stored data on internal and segregated servers that were accessible to groups of employees. There was a distinct absence of link between IT and Operational Technology (OT).

Prior to 4IR, OT was responsible solely for measurement and specifically the manufacturing process, from some raw material enters the factory till the final product is ready to be dispatched. The OT architecture comprised the hardware and software components of PLCs, DCS, SCADA, and MES systems. These systems operated in isolation from the IT systems. OT data storage was limited to confined systems, allowing access to only the factory floor. Hence, OT systems functioned in complete isolation from the organisational IT infrastructure. This resulted in a gap between production statistics, engineering data, and administrative functions. This gap created the need to integrate IT and OT, in the period of 4IR.

In the 4IR space, the integration of IT and OT has become critical in a factory environment, to synchronise and include the role of every department and employee that is a component in producing the product of the manufacturing process.

An integrated IT and OT systems create a uniform, data-driven system that bridges the gap between digital and physical operations. The data from OT devices like process measuring instruments, control systems, and final correcting elements can be linked directly into IT systems. For example, AI-enabled sensors on the factory floor can monitor the state of a process and transmit the data to an IT system, where analysis takes place through a ML algorithm to detect patterns and determine maintenance schedules. Another example is taking the data of capital human resource and salaries to synchronise this with data of production and profit. This interaction progresses to a point where IT systems basically process data whilst OT systems execute automated plant behaviour [8].

5.2 The Impact on Society and the Workforce

The development into the phase of AI, ML, and 4IR does not only benefit industry but also restructure the workforce and society in a whole. As technology replaces more manual tasks performed by humans, there is a perception that job losses will follow. However, 4IR creates new employment opportunities, by upskilling the workforce, with other added benefits as below [9]:

- **Job transformation**: Many day-to-day repetitive tasks tend to create a level of dissatisfaction in the workforce. These tasks are converted to being automated, thereby creating a work environment of a more complex nature. This eliminates any possibility of boredom. Jobs are transformed to require more critical thinking and problem-solving ability, thereby increasing the level of job satisfaction.
- **Skills development**: It is impossible for the workforce pre-4IR to be able to perform duties as expected in post-4IR. The technological requirements will require a higher level of skills, more at the lower end. Hence employees will require a minimum level of capacity development to acquire the expected skills. This becomes beneficial to both employer and employer.
- **Qualified personnel**: Whilst the required skills will be essential, the technological requirements will require a higher level of qualifications, more at the higher end. The cognitive levels of thinking and interpreting, especially with design procedures. Hence employees will require a minimum level of qualification to acquire the expected knowledge and to perform specialised work. This becomes beneficial to both employer and employee.
- **Ethical and social considerations**: The algorithms used in AI and ML can create ethical challenges, in terms of privacy and bias, amongst others. Since AI and ML have the ability to bridge the gap between human capabilities and machine operations. AI can monitor employee performance in the workplace. This creates a challenge of the level of surveillance and data privacy.

5.3 Machine Learning Techniques in Process Automation

A ML technique is an algorithm processed in a computer system to learn patterns, make predictions, and improve the accuracy of the stored data in the absence of programming each individual task. These techniques facilitate automated decision-making, from archived data. The following are ML techniques adopted in the field of process automation and control, with an application for each one:

- **Supervised learning** is a technique to learn from trained models, whereby the system learns input-output mappings, to enable predictions from stored patterns in training data.

  - *Application*: Predictive maintenance is accomplished by analysing categorised historical data on equipment failures, to predict when components within an instrument can fail, allowing for maintenance scheduling.
- **Unsupervised learning** works with data that has no known labels, to identify underlying structures in the stored data and to identify unknown patterns.
  - *Application*: Anomaly detection in real-time monitoring systems used to detect unfamiliar patterns in sensor analysis which can alert the system.
- **Reinforcement Learning (RL)** entails a learning method by trial and error, to provide feedback.
  - *Application*: Control optimisation is achieved in advanced systems, like pH control, where the RL model learns to adjust control parameters in real time.
- **Semi-supervised learning** uses a smaller amount of known data and a larger amount of unknown data to enhance model performance when known data is not available.
  - *Application*: Quality control monitoring is improved, when a smaller amount of data is known of existing defects. Model accuracy is improved in identifying faulty parts with limited data.
- **Transfer learning** works within two domains, by taking a pre-trained model from one domain, compare it to a model in the other domain, to fine-tune it and generate a model for a different domain.
  - *Application*: Predictive quality analysis for a new product can be derived from the generated model in the new domain.
- **Deep learning** accesses multi-layered neural networks to identify complex patterns in high-dimensional data and is used mainly for images and audio data.
  - *Application*: Image-based inspection in rapid prototyping applications, where models analyse images captured from cameras, e.g. labels on bottles, against the stored data.
- **Clustering**, as the name indicates, clusters groups of data points based on similarity, to compare, in the absence of predefined labels.
  - *Application*: Process parameter grouping to optimise the process, where group clusters containing different operating conditions received from field sensor data, assists operators to follow the dynamics of the operations, and be able to optimise accordingly.
- **Dimensionality reduction** streamlines data in accordance with priority of essential data, facilitating computational efficiency of the model being generated.
  - *Application*: Multi-sensor data fusion, especially in analysers, is where data from multiple sensors in the measuring and reference sensors are compressed

to essential features. Whilst allowing for faster responses without losing critical data, the secondary data remain available if required.

- **Support vector machine** creates catalogues that separates data into groups/classes.
  - *Application*: Fault classification in the manufacturing process classifies different fault types based on real-time readings from sensors, to assist operators to respond appropriately and in the level of priority, to types of failures identified.
- **Neural networks** use hierarchical layers of neurons to model complex links in data. The model can be customised for specific tasks by adjusting the network's structure.
  - *Application*: In a water treatment plant, where pH is a complex variable to control, neural networks can predict pH changes based on sensor inputs, enabling more accurate pH control.

5.4 The Future of Industry 4.0

The future of 4IR in the field of process automation and control can be characterised by some key trends and developments expected to shape the field, as follows:

- **Enhanced connectivity** will enable sensors, control systems, final correcting elements, and machinery to communicate in real time, allowing for improved monitoring and control of industrial processes, thereby reducing downtime.
- **Predictive maintenance** will facilitate planned and predictive maintenance schedules by analysing data from sensors and equipment, e.g. vibration monitors to detect factors like metal fatigue, amongst others, to identify possible failures before they occur. This reduces maintenance costs drastically and avoids much production interruptions.
- **Digital twin** creates a platform where virtual models of physical items are created, to be used in computer simulations for future planning, instead of investing in real-time systems. After required analysis and desired results are obtained, more informed decisions can be made thereafter.
- **Data-driven decision-making** will create a platform for data to be accessed for critical decision-making in the process automation environment. Factors like human resources, measurement and control technology, finance, amongst others will be considered to facilitate businesses to gain access to data into operational performance to optimise processes, operations, profits, etc.
- **Autonomous systems**, not limited to drones and autonomous vehicles, will transform areas like logistics, production, and operations in the manufacturing environment. These systems will function independently, eliminating the presence of the human. In an industry like mining, it becomes necessary for inspection

or specialised functions in hazardous areas. This is where this system becomes the only option.

- **Cybersecurity enhancement** takes cybersecurity to a higher level due to industrial systems being connected to IT systems, to deal with cyberthreats. Advanced protection including AI-driven threat recognition will be fundamental to protect sensitive data and maintain the integrity of various subsystems.
- **Energy efficiency and sustainability** will create more sustainable measures in process automation. Focus will be on energy consumption and lower environmental impact, by creating smart grids, energy-efficient equipment, and process automation and control systems that incorporate environmental control at the same level as process control.
- **Workforce transformation** will ensure that capacity is built in the workforce to possess the required skills and qualifications to embark more effectively in this new world of work. This paradigm shift in process automation and advanced technologies will shift the workforce landscape. Whilst many jobs may become redundant, new roles will be required for employees to adapt to technological advancements.
- **Collaborative robots (Cobots)** will bridge the gap between human and robots to perform specialised functions where they become dependent on each other to complete a task. Cobots possess built-in sensors with integrated safety features to work with humans, e.g. in handling dangerous chemicals in certain cases, enabling a safe and streamlined workflow, whilst providing real-time data at the same time for various reasons.
- **Cloud computing and edge computing** in a conductivity control system enable unified data storage to facilitate remote monitoring of the real-time measurements and fine-tuning of controller, to maintain conductivity levels. This methodology improves predictive maintenance by preventing any degradation of the inside of the process lines, using accumulated data to recognise trends and potential concerns. Edge computing processes the actual conductivity measurements and stores the data locally to enable the required response from the control algorithm and to maintain optimal conductivity levels.

6 Application of 4IR in Process Automation and Control

The application of AI and 4IR technology is not only confined to the industrial environment, but also equally important in the education sector. As such, the technology has been implemented in Mangosuthu University of Technology (MUT), Umlazi, KwaZulu Natal, South Africa for teaching and research purposes. The concepts of 4IR are covered in undergraduate studies up to final year projects, and research utilises fully equipped facilities at post graduate level. Past research used Siemens FuzzyControl++, industrial AI tool, with Step7 software and S7-300 hardware. This section discusses how Siemens technology evolved from the Step7, Version 5.2 to Step7, Version 13 and 14 which is 4IR compliant, including its SIMATIC controllers. The software, Totally Integrated Automation (TIA) Portal is used with 1200 PLC hardware at MUT. Additional components that can be used are Industrial Edge and MindSphere platform, to further transform process automation and control. Some of the concepts are discussed below, using Siemens 4IR technology [10]:

- **SIMATIC PLC** hardware comprise the S7-1200, S7-1400, and S7-1500 series, which are high-performance controllers making provision for various options to communicate with the programming unit, other PLCs, other IoT devices, and the process field devices. These communication options are traditional transmission signals, Profibus, Ethernet, and MultiPoint Interface (MPI) [11].
- **TIA Portal** software is an integrated engineering structure that makes provision for device programming, simulation, and data management. Programming is accomplished in Step7, to enable unified integration with other 4IR components. Simulation is accomplished through PLC-SIM, a virtual PLC.
- **Siemens industrial edge** makes provision for real-time edge computing into the industrial environment by supporting data to be processed close to the equipment. This facilitates a strong, reliable network connection for effective data analysis and efficient real-time control dynamics based on AI algorithms.
- **Siemens MindSphere** is an open source IoT operating system that interfaces production data to the cloud, making provision for advanced systematic computational analysis of data and real-time monitoring. This level of connectivity facilitates decision-making processes dependent on data and remote monitoring functions.
- **Predictive maintenance** is made possible via Siemens PLC's and TIA Portal that collect data from field measuring and final correcting instruments which then transmits it to MindSphere or an industrial edge device. Analysis of this data is processed through an AI set of rules by identifying patterns that indicate abnormal conditions on the instruments.
- **Real-time optimisation** allows systems to dynamically adapt to real-time data. Data is collected via Siemens hardware PLCs through to TIA software Portal, which then processes this vast amount of data. AI algorithms running on Siemens Industrial Edge, through these systems, recognise optimal values for process variables to increase product quality and yield production targets.
- **Energy management** is core in the manufacturing industry with the aim to reduce costs and achieve production targets. Siemens 4IR-enabled solutions allow for field data to be collected by the PLC and managed through TIA Portal which are then analysed in MindSphere to detect anomalies in energy consumption. AI algorithms analyse consumption patterns and regulate operational equipment to adjust energy consumption during peak times.
- **Enhanced quality control** is achieved via PLC that collects real-time software data on production parameters, for processing through AI and ML models that analyse this data for comparison with regional, national, and international quality standards. This is achieved via TIA Portal, through to industrial edge devices then to AI algorithms that communicate directly with the production line to immediately detect products not meeting required quality standards.

- **Autonomous operations** are achieved through Siemens PLC via Industrial Edge and MindSphere to facilitate remote and autonomous control of equipment. Siemens automation, IoT technologies, and HMI allow operators to monitor and control processes from control rooms or dedicated panels in the plant, improving production statistics, amongst other distinct benefits.
- **Enhanced flexibility and scalability** are achieved through Siemens open, modular methodology by incorporating new field devices or updated software applications. The scalability of SIMATIC PLC to Step7 to MindSphere allows businesses to customise automation systems to their specifications. This flexibility facilitates the company's ability to adapt to market and technology demands.
- **Data analytics** is driven by Siemens Industrial Edge and MindSphere platforms to provide data to all departments in the manufacturing process for proactive decision-making. Real-time data is accessible operators and managers to attend to production and maintenance issues if they arise. AI algorithms are integral in the system design within the Siemens architecture for predictive and prescriptive factors, to ultimately optimise the operational processes.
- **Enhanced sustainability and compliance** are achieved through Siemens 4IR-aligned process automation systems to enable companies in sustaining their operational standards, at the same time complying with requirements of regulatory bodies. The technology used maintains both control and environmental goals.

7 Benefits of 4IR, AI and ML in Education

4IR, AI and ML is transforming all sectors of life, including our lifestyle. Apart from this new technology benefitting industry, it is essential for this new way of living to commence in the teaching and learning phases. Ideally, it needs to start at early childhood level all the way through to university studies. Currently, as this new technology has suddenly appeared, there are different categories of human experiences. A certain percentage of people will be part of this technology in different sectors of engineering, technology, and productions, amongst others. Integrating 4IR, AI and ML, into education is an integral component of preparing students to for this technology-driven world. By teaching these skills and knowledge, educational institutions adopt an approach of preparing future engineering practitioners with the required skills, who will be able to think critically and embrace this new future confidently, by providing the required student tools, empowering them to be proactive participants in this evolutionary paradigm. This section will discuss some highlighted benefits of teaching 4IR technologies like AI and ML and briefly support each point with proposed examples to emphasise the benefits [12].

7.1 Developing Problem-Solving and Critical Thinking Skills

It is a known fact that daily in the engineering workspace, the level of competency may not really be evaluated. When a breakdown occurs, the level of competency will determine the time it takes to trouble-shoot, fault-find, and resolve the situation, dependent on the level of critical thinking to analyse and action. AI and ML are on a level that works with complex problems, thereby expecting students to analyse data and make informed decisions. Teaching these skills in education becomes imperative to promote the level of critical thinking and build on the skills required to analyse problem situations to develop suitable solutions [13].

An example could be distributing data sets to students and expecting them to design ML models to predict specific outcomes. A simple program could be written on a micro controller for ambient temperature that varies daily.

7.2 Preparing Students for Future Job Markets

Very often the reality is that knowledge and skills acquired at the exit level of an engineering program may not necessarily be required by the time a graduate enters the workplace. Hence, AI and ML skills are progressively in demand, as industries adapt with these technologies. Introducing 4IR, AI and ML, in the classroom builds capacity in students to acquire the expected skills for the respective job market, allowing them to adapt to this paradigm. Understanding AI concepts makes students more marketable, improving their employability options [14].

An example could be offering AI courses within the existing curriculum to build capacity in students to analyse data, construct ML algorithms, and derive output solutions.

7.3 Building Innovation and Entrepreneurship

Most often graduates choose to seek employment instead of becoming an entrepreneur. With this approach, it is evident that students may lack a sense of innovation. Including AI and ML concepts in the curriculum teaches students to use their sense of imagination, to build on innovation in them. This sense of innovation encourages them to think differently, think creatively and think business, considering entrepreneurial opportunities. In the student world, one is used to making applications, from applying for a place at school or college or university. Hence it becomes a norm to apply for a job as a graduate. This way of thinking can only change with a different skill set. The concepts of AI and ML encourage engineering students to analyse market trends and market needs, especially in this era of electricity and water challenges. This approach will steer students in the direction of seeking opportunities in business, employing people instead of seeking employment and becoming an employee. 4IR technologies empower students to become inventors instead of becoming other consumers of the technology [15].

An example could get students to develop applications that leverage AI to influence some change in society, such as water conservation. As we are aware that water cannot be created in another form like electricity, the only option is to conserve water. An AI-driven app could consider how any leak on a waterline, anywhere, could be detected, to arrange shutting off that line and preventing wastage of water immediately, until it is repaired.

7.4 Personalised Learning and Adaptive Education

Rote learning is very common in the teaching and learning process whereby students cram knowledge for an assessment in a specific time limit. The outcome from this method may not necessarily reflect competency. AI-aligned educational tools provide customised learning experiences, adapting to students' learning pace and technique. AI concepts teach students to build a sense of creativity, thereby adapting to an appropriate learning platform that they adjust to. This approach increases their level of understanding of AI-related applications and how they function in the real world.

An example can be creating an adaptive learning platform in Maths education. Mathematics is generally a subject that creates challenges to students. Hence AI algorithms could be generated to personalise each lesson to allow each student to adapt to a different set of knowledge-based rules, depending on their individual performance and progress. This method is sure to improve the level of students' math competence since students will interact with their study material that is customised to their learning rate. This method will also expose students on how adaptive algorithms function, the same principle that lies behind the technology used [16].

7.5 Promoting Data Literacy and Ethical Understanding

Accessing data with integrity and adopting an ethical approach are fundamental components in the AI and ML space. Data literacy is critical in the IT world where 4IR, AI and ML, resides. Hence this type of education must teach students what integrity and ethics really mean in this technological world, especially some data may infringe on privacy very often, in various ways. Teaching AI and ML makes it compulsory to engage in discussions on ethical considerations, how to deal with data privacy, how to ensure that the software solutions take into consideration algorithmic bias, and the impact on society at large. The ethical aspects of AI teach students to adopt a responsible attitude toward using the appropriate technology and how to implement it without any compromise.

An example could be an exercise where students conduct case studies on facial recognition technology with their fellow students, at the same time ensuring that they are aware of privacy concerns and potential biases in algorithms. The intention is to include a component of ethics in appropriate courses, to help create the awareness of how the new technology could make compromises, if not implemented with integrity and ethics. This will prevent students losing oversight on integrity and ethics while focusing on core learning areas of AI and ML [17].

7.6 Establishing Collaboration and Interdisciplinary Skills

In the 4IR space, AI and ML will most certainly overlap various disciplines, from education to engineering to accounting to medicine to knowledge amongst many others. Whilst students will cover concepts of AI and ML in their respective programs, the application of the knowledge and skills will overlap another discipline. As such it becomes essential that students function in collaborative environment, having some knowledge and skills of the other disciplines, whilst gaining full understanding of their specific field. Collaboration adopts teamwork, improves communication skills, and develops project management skills.

An example could be a collaborative engineering project to design and commission a pilot batching plant that includes concepts of electrical, mechanical, chemical, civil, computers, and IT. A project of this magnitude is sure to facilitate students gaining experience in interdisciplinary collaboration, especially in the AI space.

7.7 Encouraging Lifelong Learning and Adaptability

Since inception of 4IR, following AI and ML, it is sure to be a progressive path as development in these evolving fields are sure to continue. As such the learning experience by students must become a lifelong path as well, to ensure that they can adapt to the environment at various stages in their career. Students are bound to be faced with new challenges, thereby compelling them to maintain the pace of technological advancements. It will be expected of students to develop the flexibility in the way they work, to adapt to the challenges. As such, these qualities must be covered in the teaching and learning phases, ensuring that students remain competitive throughout their careers.

An example could be allocating a project that is based on classroom work but can be adapted into being used at home. Applying classroom knowledge and skills at home is core in building understanding in a student. Reason being that it is at home that students apply their minds daily. Hence, allocating a simple project based on the principle of day-night sensing can be used in various ways at home. Students must grow with the understanding that AI is applicable universally, encouraging them to explore into this developing world of technologies [18].

7.8 Enhancing Engagement and Interest in STEM Fields

The fundamentals are taught in the Science, Technology, Engineering, and Maths (STEM) fields to embark on this journey into 4IR, AI and ML. Hence it becomes imperative for students to engage and develop an interest in STEM. We have passed the stage where a portion of the population was not computer literate. However, we have all learnt that being computer literate became compulsory and is now part of our daily lives. The cell phone is such an example where we interact with apps that are AI-driven. As such, students need to relate with STEM concepts through interactive physical projects, software programming, and simulations. In this way, AI and ML education will change the abstract understanding of STEM principles to a concept that is more meaningful. Students must be taught to solve real-world challenges instead of retaining their knowledge within the cover of their books and their minds. In this way, they will be encouraged to pursue careers in STEM fields and perhaps into AI and ML.

An example is to include an elective subject covering basics of 4IR, AI or ML, to create awareness and encourage interest in STEM careers. Going further, students should engage with interactive projects like developing chatbots since all fields of study have a component of using the principles of flow charts. At times, students use it but are unaware of the principles. It's like making a cup of coffee that is based on this principle. Hence, all courses should encourage students to pursue studies and careers in AI and ML technology, to contribute to building capacity in these fields.

7.9 Preparing for Responsible Citizenship in an AI-Driven World

The fundamentals taught in an AI or ML study program cover aspects of integrity and ethics. However, AI becomes more invasive in daily life; thereby, understanding its implications becomes fundamental to be a responsible citizen. Educating students on AI and ML principles prepares them to become responsible citizens as they become fully aware on how to critically assess AI-driven systems and understand the impact of automation and machine intelligence, to the level of society. Very often we may not be aware of wrongdoing when using certain apps which can be avoided by following an AI-driven study route.

An example is for institutions to offer an online AI program, open to the public, as a short learning program. In this way students gain a basic understanding of how AI functions in our lives and the ethical factors surrounding it, encouraging them to interact in the public space in an appropriate manner, especially around AI-related matters.

8 Challenges and Considerations

With the phenomenally fast pace at which technology is changing, resulting in the benefits of implementing 4IR, AI and ML, in industry and teaching, it does not come without challenges. Some of the challenges are related to data security, system integration, workforce readiness, ethical considerations, and implementation of the technologies. Identifying and addressing these challenges are essential to ensuring that AI and ML technologies generate the expected outcomes in process automation and control. This section will discuss the challenges and possible considerations.

8.1 Data Security and Privacy

AI and ML systems depend on substantial volumes of reliable, high-quality data to design the appropriate solution. Due to systems being vulnerable to cyberthreats, various challenges need to be considered to ensure that the availability, accuracy, security, and integrity of data are not compromised, especially in organisations where the sources of data are disconnected with each other. Especially in the process automation and control environment, AI and ML systems deal with the continuous collection and analysis of large amounts of measurement and operational data. This data can include complex production and sensor information that can become vulnerable to cyber-attacks or unauthorised access. Data security and privacy are therefore one of the most critical challenges in 4IR applications. A violation in data security could result in production losses, financial consequences, and compromised safety of equipment or people.

8.1.1 Considerations

- **Encryption and access control**: Encoding equipment is essential. Access control mechanisms to monitor and restrict unauthorised access.
- **Real-time monitoring**: Real-time AI-driven cybersecurity systems need to monitor the network traffic to detect and flag potential threats in advance, for possible intervention.
- **Compliance with regulations**: Compliance with data privacy standards and regulations regionally, nationally, and internationally are essential, to warrant responsible and ethical processing of data that is collected from the automation processes.

8.2 *Integrating AI and ML with Existing Systems*

The technology of existing systems as compared to AI and ML based technology creates another significant challenge. Whilst industries need to adapt to maintain a place in the market, this changeover becomes complex. Existing systems may not support modern data processing and connectivity, especially in the way the existing communication structure needs to adapt to this new topology. From the field instruments to the automation and control hardware, adaptability becomes a major challenge due to the facilities not being compatible with IIoT or data-driven AI applications. This changeover can be costly, technically demanding and disruptive to production.

8.2.1 Considerations

- **Hybrid integration solutions**: Consideration is given to install hybrid models where existing systems are integrated using edge computing devices to serve as interface to collect and process data locally with limited AI functionality, to avoid a complete system installation.
- **Incremental upgrades**: Another option is to install AI and ML in phases, to upgrade by priority, focusing on critical components in the absence of a complete replacement. Priority is given to areas like predictive maintenance, to generate instant benefits while gradually planning the rest of the upgrades.
- **Standardised protocols**: Open Platform Communications Unified Architecture (OPC UA) is one such 4IR compliant protocol that enables compatibility between old and new systems, for a changeover with minimal or no disruptions. OPC UA ensures secure and reliable data exchange between devices and systems.

8.3 *Workforce Skills Gap*

Having a workforce that has adapted and mastered the way of working with existing technology creates a challenge to change this new technology. The implementation of AI and ML requires specialised skills and knowledge. Very often there is a significant skills gap where the existing workforce requires some form of capacity building to function in this new world of work. This skills gap can be phenomenal and may result in reduced returns on investment in 4IR technologies [19].

8.3.1 Considerations

- **Training programs**: Training programs in AI and ML will become essential as there is no other option for new and existing employees, to bridge the skills gap and build capacity to empower the workforce to manage the technical demands of 4IR applications. This serves well at the level of artisans.
- **Partnerships with educational institutions**: Collaboration with educational institutions will become another essential initiative to develop 4IR-based curriculum to meet industry requirements. This new curriculum will ensure that graduates are professionally qualified to meet the demands of modern 4IR space in process automation and control. This serves well at the level of technicians and engineers [20].
- **User-friendly AI solutions**: Implementing AI and ML tools with insightful interfaces with platforms that facilitate effective interaction between man and machine, avoiding the need for extensive training. Whilst training is required to build capacity amongst the workforce, the factor of time becomes a concern since members of the workforce will be absent from their workplace during training periods.

8.4 Data Quality and Availability

The quality and continuous availability of data is an integral component for an AI and ML system to function optimally. As such these systems depend on massive amounts of high-quality data to yield expected results. In the industrial environment, data quality can be compromised in many ways, e.g. ambient conditions, faulty sensor, incorrect transmission signal, data collection processes not representative of signals, etc. Compromised data can prejudice the accuracy of AI models, leading to compromised decisions, resulting in errors in process control or safety risks. This compromised data results in the historical data generating an incorrect model, for training the system.

8.4.1 Considerations

- **Data cleaning and preprocessing**: Data cleaning and preprocessing protocols are mechanisms to ensure that compromised data are further processed to be fed into AI and ML models, to ensure that they are consistent and accurate. Automated data validation systems are part of this process to detect anomalies to ensure that only reliable data is used.
- **Synthetic data for model training**: When historical data is insufficient, artificial data is generated to make up for the shortfall, which assists in training AI models by generate credible operational circumstances. This method helps AI models train to avoid expecting extensive historical records.

- **Continuous data quality monitoring**: Continuous monitoring of data quality is essential to allow for early detection of anomalies, ensuring that AI models are accurate.

8.5 *Ethical and Regulatory Considerations*

Abiding by codes of ethics and complying with legislation contributes to a reliable AI system. The implementation of AI and ML in process automation introduces the challenge of ethical and regulatory factors. This includes the privacy and transparency of data to establish the level of accountability, using the data. Automated decisions based on human reasoning from a machine, generated from a trained AI-driven control algorithm introduce ethical challenges, specifically in applications where safety is of significance. Equally important is compliance with regional, national, and international legislation, to avoid legal and financial implications [21].

8.5.1 Considerations

- **Transparency and explainability**: AI models need to be built to enable operators to understand the logic behind intelligent decision-making rules. Transparency in AI algorithms encourages confidence and facilitates practical use of the architecture by the operators.
- **Adherence to ethical standards**: Compliance with industry-specific standards results in codes of ethics being followed, with the prescribed guidelines to ensure that AI applications abide by privacy and ethical standards. This becomes critical in sectors like the food industry, where AI may have safety implications.
- **Regular audits and compliance monitoring**: Regular audits of AI and ML systems certify adherence to legal requirements and ethical norms. Compliance monitoring ensures that accountability and transparency in the use of technologies are in place.

8.6 *Model Accuracy and Adaptability*

Accuracy and adaptability of the designed algorithms are paramount to ensure that process requirements are met. AI and ML models employed in process automation runs on a continuous basis, monitoring process conditions and regulating final correcting elements, allowing the process to adapt to changing conditions. Most industrial environments are extremely dynamic with variables like pH being the measured variable, at the same time being affected by secondary variables like temperature, flow, and pressure, thereby shifting desired values and affecting model performance intermittently. An effective AI model may lose accuracy over time, compromising control conditions.

8.6.1 Considerations

- **Model retraining and updating**: Continuous model retraining can be set up using recent data to optimise accuracy for the abnormal process conditions. Automated model retraining entails using the various process conditions to develop models to adapt to these changing conditions without requiring manual interference.
- **Performance monitoring tools**: Performance monitoring tools are implemented for real-time measurements to gauge model accuracy, making provision for alarms when models require retraining or instruments require recalibration. These tools contribute to preventing disruptions in the process and to maintain output requirements.
- **Simulation and testing environments**: Simulating the AI model is accomplished in a virtual environment with inputs, outputs, and the architecture being tested offline. When the simulated results on performance of the designed model is tested under different conditions, and the expected results are obtained, the model is commissioned in real time on the physical plant to conduct endurance tests.

9 Conclusion

The process automation and control landscape has been transformed by 4IR technology, supported by AI and ML. This development has further developed traditional manufacturing processes by integrating advanced software and hardware components, improving factors like precision of equipment, efficiency of operations, and optimised automation systems. AI and ML within the 4IR framework have created the ability to analyse substantial amounts of data generated by IIoT devices. AI-structured algorithms can predict anomalies before they occur, initiating corrective system actions. Similarly, ML models develop control strategies by learning from archived plant data, enabling them to adapt to changing process conditions without human intervention. The application of AI and ML strategies in process control optimises the functional work of operations and technical, at all levels of industrial operations. SCADA, implemented with intelligent algorithms, can independently optimise complex processes. HMIs have become more perceptive, functioning as an intelligent device, where operators can interact with systems with voice and other forms of recognition [22].

Concerns about cyberthreats, workforce implications, and the adapting intelligent systems into existing systems create a different set of challenges. Implementing robust multi-layered security frameworks and advanced encryption protocols can contribute to mitigate risks with cyberthreats. The workforce can be prepared for AI and ML systems through customised upskilling and reskilling programs. Adapting AI and ML systems into existing infrastructure entails modifying existing systems with compatible technologies to ensure seamless switch over and minimal interruption to existing operations.

The evolution of process automation and control in the 4IR era, with AI and ML technologies, has initiated a gigantic surge towards smarter and sustainable IT and OT practices. These technologies will unlock exceptional prospects for further revolution, ensuring that industries remain robust and competitive, globally. The interaction between human proficiency and machine learning will form the foundation of this revolution, modelling a new future for process automation and control.

References

1. Velarde, G.: Artificial intelligence and its impact on the fourth industrial revolution: a review. Int. J. Artif. Intell. Appl. **10**(6), 41–48 (2019)
2. Schwab, K.: The Fourth Industrial Revolution (2016). www.weforum.org
3. Herweijer, C., Waughrey, D.: Fourth Industrial Revolution for the Earth Harnessing Artificial Intelligence for the Earth (2018)
4. Sutcliffe, M.: RFQ-4th Industrial Research/08/572 Local Government Sector Education and Training Authority (LGSETA) (2020). www.cityinsight.co.za
5. Van Vuuren, G.J.: An Introduction to Process Instrumentation. Quad Technologies, 184 p. (2010)
6. Naidoo, P.: The Control of a Multi-Variable Industrial Process, by Means of Intelligent Technology (2001)
7. Naidoo, P.: Intelligent Control and Tracking of a Solar Parabolic Trough (2006)
8. Bin, R.A., Kausik, M.A.K.: AI revolutionizing industries worldwide: a comprehensive overview of its diverse applications. Hybrid Adv. **7**, 100277 (2024)
9. Xu, M., David, J.M., Kim, S.H.: The fourth industrial revolution: opportunities and challenges. Int. J. Financ. Res. **9**(2), 90–95 (2018)
10. Naidoo, P., Sibanda, M.: Integrating artificial intelligence in electrical engineering education: a framework for the fourth industrial revolution. In: Focus Conference (2024)
11. French, A.M., Shim, J.P., Larsen, K.R., Risius, M., Jain, H.: The 4th industrial revolution powered by the integration of AI, blockchain, and 5g. Commun. Assoc. Inf. Syst. **49**, 266–286 (2021)
12. Naidoo, P., Sibanda, M.: Emerging trends and future directions of the industrial internet of things. In: From Internet of Things to Internet of Intelligence Transactions on Computational Science and Computational Intelligence, pp. 91–110. Springer (2023)
13. Rizk, N., Ismail, A.: The Fourth Industrial Revolution, Artificial Intelligence, and the Future of Work in Egypt. International Labor Organization (ILO) (2020)
14. Oosthuizen, R.M.: The fourth industrial revolution—smart technology, artificial intelligence, robotics and algorithms: industrial psychologists in future workplaces. Front. Artif. Intell. **5** (2022). https://doi.org/10.3389/frai.2022.913168
15. Buyana, N.: The Effect of the Fourth Industrial Revolution on Employment in the Insurance Sector in Cape Town, South Africa (2022)
16. Ally, M., Perris, K.: Artificial intelligence in the fourth industrial revolution to educate for sustainable development (L'intelligence artificielle dans la quatrième révolution industrielle

pour éduquer au développement durable). Can. J. Learn. Technol. **48** (2022). https://doi.org/10.21432/cjlt28287
17. Malik, S., Muhammad, K., Waheed, Y.: Artificial intelligence and industrial applications—a revolution in modern industries. Ain Shams Eng. J. **15**, 102886 (2024)
18. Ibegbulam, C., Olowonubi, J.A., Fatounde, S.A., Oyegunwa, O.A.: Artificial intelligence in the era of 4IR: drivers, challenges and opportunities. Eng. Sci. Technol. J. **4**(6), 473–488 (2023). www.fepbl.com/index.php/estj
19. Naidoo, P., Sibanda, M.: Building an entrepreneurial culture in electrical engineering education. In: Focus Conference (2023)
20. Baijnath, N.: Sarchi Teaching and Learning Research Paper Series (October 2021)
21. Naidoo, P., Sibanda, M.: Embedded supercomputing for edge devices: a comprehensive review. In: From Internet of Things to Internet of Intelligence Transactions on Computational Science and Computational Intelligence, pp. 157–173. Springer (2023)
22. Ribeiro, J., Lima, R., Eckhardt, T., Paiva, S.: Robotic process automation and artificial intelligence in Industry 4.0—a literature review. Proc. Comput. Sci. **181**, 51–58 (2021)

Machine Learning for Predictive Modeling of Co-gasification: A Step Toward Industry 4.0 in Biomass Waste Valorization

Ocident Bongomin, Charles Nzila, Josphat Igadwa Mwasiagi, and Obadiah Maube

1 Introduction

The convergence of Industry 4.0 and biomass waste valorization represents a pivotal advancement in sustainable energy production and waste management [1]. Industry 4.0 revolutionizes industrial processes through intelligent automation and data-driven decision-making, offering unprecedented opportunities to enhance the efficiency and effectiveness of various technologies [2–5]. Among these, co-gasification emerges as a promising method for converting biomass waste into valuable syngas, a crucial step toward achieving a circular economy [6].

Co-gasification involves the simultaneous conversion of multiple feedstock, typically a mixture of biomass and fossil fuels or other waste materials, into syngas [7]. This process leverages the partial oxidation of the feedstocks under high-temperature conditions in the presence of a gasifying agent, such as steam or air. By optimizing the combination of different feedstocks, co-gasification enhances syngas quality

O. Bongomin (✉)
Department of Manufacturing, Textile and Industrial Engineering, Moi University, School of Engineering, Eldoret, Kenya

Center of Excellence in Phytochemical, Textile and Renewable Energy, Moi University, Eldoret, Kenya
e-mail: ocident@mu.ac.ke

C. Nzila · J. I. Mwasiagi
Department of Manufacturing, Textile and Industrial Engineering, Moi University, School of Engineering, Eldoret, Kenya
e-mail: charlesnzila@mu.ac.ke

O. Maube
School of Mechanical and Manufacturing Engineering, Faculty of Engineering and Built Environment, Technical University of Kenya, Nairobi, Kenya

P. Naidoo, M. Sibanda (eds.), *Measurement, Automation, and Control in Artificial Intelligence and Machine Learning*, Signals and Communication Technology, https://doi.org/10.1007/978-3-031-91802-5_5

and yield, contributing to improved carbon conversion and reduced tar formation [8, 9]. The resulting syngas, comprising hydrogen (H_2), carbon monoxide (CO), carbon dioxide (CO_2), and methane (CH_4), serves as a versatile fuel for energy production and a valuable feedstock for chemical synthesis.

Biomass waste valorization is integral to sustainable development, addressing waste management challenges and promoting renewable energy sources [6]. The efficient utilization of agricultural residues, forestry by-products, and municipal solid waste (MSW) can significantly reduce greenhouse gas emissions, mitigate environmental pollution, and enhance energy security [10–12]. The valorization of biomass waste not only fosters economic opportunities, particularly in rural areas, but also aligns with global sustainable development goals, such as affordable and clean energy (SDG 7) [13].

In the context of Industry 4.0, integrating machine learning (ML) techniques with co-gasification processes presents a transformative approach to predictive modeling [14]. ML algorithms, including Artificial Neural Networks (ANN), Random Forest (RF), Gaussian Process Regression (GPR), and Gradient Boosting Regression (GBR), offer powerful tools for predicting syngas composition and optimizing process parameters [15]. By harnessing real-time data acquisition and intelligent algorithms, ML can continuously improve model prediction accuracy, enabling optimized process control strategies and maximizing the efficiency of biomass waste valorization [16].

This study explores the potential of ML techniques for predictive modeling of co-gasification, aiming to enhance the integration of Industry 4.0 principles in biomass waste management. Through a comprehensive analysis of experimental data obtained from the co-gasification of various biomass feedstocks, the efficacy of ANN, GPR, and GBR in predicting syngas composition was evaluated. The findings demonstrate the superiority of GBR, with an R^2 exceeding 0.85, indicating highly accurate prediction capabilities. By leveraging these advanced ML models, this research contributes to the advancement of Industry 4.0 in the biomass waste management sector, promoting sustainable energy production and efficient waste utilization.

2 Literature Review

The literature on co-gasification and biomass waste valorization has expanded significantly in recent years, reflecting the growing importance of these technologies in renewable energy production and sustainable waste management. Co-gasification, which involves the simultaneous gasification of multiple feedstocks, has emerged as a promising approach for producing high-quality syngas, enhancing the efficiency of the gasification process, and enabling the effective utilization of diverse carbonaceous materials. Simultaneously, the valorization of biomass waste through various technological methods offers sustainable solutions for reducing landfill waste and generating renewable energy. Additionally, the integration of machine learning techniques in co-gasification processes demonstrates their potential to optimize syngas composition prediction and process performance, aligning with the principles of Industry 4.0. This section provides a comprehensive review of the key concepts, methods, and applications related to co-gasification, biomass waste valorization, and the role of machine learning in advancing these technologies.

2.1 Overview of Co-gasification

Co-gasification represents an innovative approach in the field of renewable energy and waste management, involving the simultaneous gasification of multiple feedstocks, such as biomass, fossil fuels, and other waste materials [7]. This process produces syngas, a versatile gas mixture primarily composed of hydrogen (H_2), carbon monoxide (CO), carbon dioxide (CO_2), and methane (CH_4). By leveraging the synergistic effects of combining different feedstocks, co-gasification can enhance the efficiency and flexibility of the gasification process, resulting in improved syngas quality and yield [17]. The principles of co-gasification, influenced by factors such as feedstock composition, gasification conditions, and the type of gasifying agent, enable the conversion of diverse carbonaceous materials into a valuable energy resource [18]. This section delves into the fundamental aspects of co-gasification, including its definition, the types of biomass feedstocks utilized, and the typical composition of the resulting syngas, highlighting its significance in advancing sustainable energy solutions and effective waste utilization.

2.1.1 Definition and Principles of Co-gasification

Co-gasification refers to the simultaneous gasification of multiple feedstocks, typically a mixture of biomass and fossil fuels or other waste materials, to produce syngas (synthesis gas) [19]. This process involves the partial oxidation of the feedstocks under high-temperature conditions in the presence of a gasifying agent, usually steam (H_2O) or air, to convert the carbonaceous materials into a combustible gas. The principle behind co-gasification lies in enhancing the overall efficiency and flexibility of the gasification process by combining different feedstocks, which can improve syngas quality and yield [20].

The process of co-gasification is influenced by several factors including the feedstock composition, gasification conditions (such as temperature and pressure), and the type of gasifying agent used [18, 21]. The interactions between different feedstocks during co-gasification can lead to synergistic effects, such as improved carbon conversion and reduced tar formation, which are beneficial for optimizing the gasification process. Additionally, co-gasification allows for the effective utilization of various waste materials, contributing to waste management and environmental sustainability [8].

2.1.2 Types of Biomass Feedstocks

Biomass feedstocks used in co-gasification can be categorized into several types, each with unique characteristics that influence the gasification process. Commonly used biomass feedstocks include agricultural residues, forestry residues, and dedicated energy crops. Agricultural residues, such as straw and corn stover, are abundant and often used due to their high availability and relatively low cost [22]. Forestry residues, including sawdust and wood chips, are also frequently utilized due to their high energy content and favorable gasification properties [23]. Dedicated energy crops, such as switchgrass and miscanthus, are cultivated specifically for energy production and offer high biomass yields and energy content [24].

The selection of biomass feedstocks for co-gasification is based on several factors including their physical and chemical properties, such as moisture content, ash content, and energy density. These properties affect the efficiency of the gasification process and the quality of the resulting syngas. The choice of feedstocks also influences the overall economic feasibility and environmental impact of the gasification system [25].

2.1.3 Typical Syngas Composition

The composition of syngas produced from co-gasification varies depending on the feedstocks used, gasification conditions, and the types of gasifying agent. Typically, syngas consists of a mixture of H_2, CO, CO_2, and CH_4, with the proportions of these components influenced by the specific feedstocks and gasification parameters such as temperature, pressure, equivalence ratio, and gasifying agents [26].

H_2 and CO are the primary components of syngas and are essential for various downstream applications, including energy production and chemical synthesis. H_2 is a clean fuel with high energy content, while CO can be used as a feedstock for chemical production [27]. The presence of CO in the syngas is a result of the incomplete conversion of carbon-containing feedstocks, and its concentration can affect the overall efficiency of the gasification process [28]. CH_4, while present in smaller quantities, can contribute to the heating value of the syngas and is typically controlled through process optimization to minimize its formation [29].

2.2 *Biomass Waste Valorization*

Biomass waste valorization is an essential component of sustainable waste management and renewable energy production [30]. By transforming organic waste materials, such as agricultural residues, forestry by-products, and MSW, into valuable resources, biomass waste valorization addresses critical environmental and economic challenges [31, 32]. This process not only reduces the volume of waste destined for landfills but also generates renewable energy and value-added products, contributing to the circular economy [33]. The integration of innovative technologies and methods for biomass conversion is pivotal in mitigating greenhouse gas emissions, promoting energy security, and fostering economic growth, particularly in rural areas where biomass resources are abundant [34, 35]. This section explores the importance, technologies, and case studies of biomass waste valorization, highlighting its significant role in achieving sustainable development goals and creating a more resilient and eco-friendly energy landscape.

2.2.1 Importance of Biomass Waste Valorization

Biomass waste valorization plays a pivotal role in sustainable energy production by converting organic waste materials into valuable resources, addressing both waste management and energy needs [36]. With the growing emphasis on sustainable development and the circular economy, efficiently utilizing biomass waste is essential for reducing greenhouse gas emissions, mitigating environmental pollution, and promoting renewable energy sources. Biomass waste, derived from agricultural residues, forestry by-products, and MSW, represents a significant and underutilized resource for bioenergy production [31].

Valorizing biomass waste reduces fossil fuel dependency and enhances energy security. It offers a renewable and carbon-neutral energy source, crucial for combating climate change and achieving global SDG 7 [37]. Additionally, converting biomass waste into energy and value-added products can create economic opportunities, particularly in rural areas where biomass waste is abundantly available, fostering job creation and economic growth.

2.2.2 Technologies and Methods for Biomass Waste Valorization

Several technologies and methods have been developed for biomass waste valorization, each with distinct advantages and applications. These include thermochemical, biochemical, and physicochemical processes.

Thermochemical processes: Thermochemical processes involve applying heat to convert biomass into energy and valuable products. Key methods include pyrolysis, gasification, and hydrothermal liquefaction. Pyrolysis involves the thermal decomposition of biomass in the absence of oxygen, producing bio-oil, biochar, and syngas [38]. This process can be optimized for different products by adjusting temperature and residence time [39]. Biomass gasification converts organic materials into syngas through partial oxidation at high temperatures. Syngas can be used for electricity generation or as a feedstock for chemical synthesis [40, 41]. Hydrothermal Liquefaction (HTL) converts wet biomass into bio-crude oil under high pressure and moderate temperature, suitable for high-moisture biomass and resulting in high-energy-density biofuels [42].

Biochemical processes: Biochemical processes utilize microorganisms and enzymes to break down biomass into biofuels and biochemicals. The primary methods are anaerobic digestion and fermentation. Anaerobic Digestion (AD) converts organic matter into biogas (CH_4 and CO) and digestate through microbial activity in the absence of oxygen. It is a well-established technology for treating various organic wastes, including agricultural residues and sewage sludge [42]. Fermentation involves the microbial conversion of sugars in biomass into ethanol and other biochemicals. This process is widely used for bioethanol production from crops like corn and sugarcane, as well as lignocellulosic biomass after pretreatment [43].

Physicochemical processes: Physicochemical processes combine physical and chemical treatments to enhance biomass conversion. Pretreatment methods, such as steam explosion, acid hydrolysis, and alkali treatment, are employed to break down lignocellulosic biomass into fermentable sugars, enhancing the efficiency of subsequent biochemical processes [44]. Chemical valorization involves converting biomass-derived intermediates into chemicals and materials, such as bio-based polymers and natural fiber-reinforced composites. These materials have applications in various industries, including packaging, construction, and automotive sectors [43].

2.2.3 Case Studies of Biomass Waste Valorization

Several successful biomass waste valorization projects demonstrate the potential and versatility of these technologies. Recent advancements in microbial engineering have led to the development of synthetic microbial consortia (SMC) for biomass waste treatment. SMC leverages synergistic interactions among different microorganisms to enhance waste degradation, control hazardous substances, and convert biomass into high-value products. For instance, Zhou et al. [41] highlighted the use of SMC in the degradation of complex waste streams, resulting in improved efficiency and stability of AD processes.

Constructed wetlands (CWs) are effective in wastewater treatment but generate significant amounts of waste biomass. Sharma et al. [43] reviewed the valorization of CW biomass into natural fiber-reinforced polymer composites (NFRPCs), which are sustainable alternatives to synthetic fiber-reinforced composites. The study emphasized the potential of using Canna indica fibers from CWs as reinforcement in polypropylene composites, showcasing a novel application of waste biomass in the polymer industry.

Tatla et al. [42] investigated the integration of HTL and AD for comprehensive biomass utilization. This integrated approach addresses the limitations of individual processes, enhancing overall efficiency and energy recovery. The study provided valuable insights into sequencing HTL and AD, optimizing bio-crude and biogas production from various biomass feedstocks.

In Brazil, large amounts of yellow passion fruit peels are generated as waste. de Castro et al. [45] explored the chemical characterization and potential applications of these peels, including pectin extraction and the production of cellulose and hemicellulose. The study highlighted the industrial relevance of valorizing passion fruit peels, contributing to waste reduction and the production of valuable biochemicals.

Another study by Umer et al. [46] demonstrated the extraction of sodium silicates (SS) from silica-rich agricultural waste ashes (AGWA) for use in geopolymer concrete. The study provided a comprehensive techno-economic analysis (TEA) and life cycle assessment (LCA), showing significant reductions in greenhouse gas emissions and costs associated with AGWA-derived SS compared to commercial SS. This innovative approach promotes circular construction and the sustainable utilization of agricultural waste.

Liu et al. [47] investigated the microwave co-pyrolysis of textile dyeing sludge and furfural residue, demonstrating enhanced bio-oil and gas production with reduced biochars. The study highlighted the potential of co-pyrolysis in mitigating environmental pollution and producing high-quality bio-products, offering new insights for the efficient utilization of industrial solid waste.

These case studies underscore the diverse applications and significant benefits of biomass waste valorization technologies, contributing to sustainable energy production, waste management, and the development of a circular economy.

2.3 Machine Learning in Co-gasification

Co-gasification, the process of converting a mixture of feedstocks into syngas, presents a promising approach for waste management, renewable energy production, and greenhouse gas mitigation. Predictive modeling plays a crucial role in optimizing co-gasification processes by enabling the prediction of syngas composition and process performance under various conditions. Machine learning (ML) techniques are increasingly employed for this purpose due to their ability to handle complex non-linear relationships between process variables and outputs.

2.3.1 Overview of Machine Learning Techniques

This section explores the application of various ML techniques in co-gasification, including support vector machine (SVM), random forest (RF), gradient boosting regression (GBR), recurrent neural network (RNN), multilayer perceptron (MLP), artificial neural network (ANN), linear regression (LR), and Gaussian process regression (GPR).

SVM is effective for high-dimensional data classification and regression problems encountered in co-gasification modeling [48]. RF is well-suited for capturing complex interactions between variables and has used for predicting H_2 yield and char formation in Chemical Looping Gasification (CLG) [48]. GBR demonstrates superior performance in predicting H_2 yield and char formation in CLG compared to SVM and RF [48].

RNN is particularly valuable for modeling dynamic systems like fluidized-bed reactors used in Supercritical Water Gasification (SCWG). RNNs can predict syngas composition and optimize process conditions for desired outputs [49]. MLP efficiently predicts gasification performance of polyethylene waste with high accuracy [50].

ANN can handle complex non-linear relationships and outperform traditional regression models in predicting various performance metrics of coal and biomass gasification processes [51]. LR is a simpler method used in conjunction with other techniques, such as correcting discrepancies between simplified models and real process data for microalgae biomass gasification [52]. GPR is a non-parametric approach suitable for small datasets and has been employed to identify optimal gasification pathways for specific biomass feedstocks [53].

2.3.2 Applications of Machine Learning for Syngas Prediction

Several studies have demonstrated the effectiveness of ML in predicting H_2 yield, char formation, and other process outputs in CLG [48]. Machine learning has been successfully applied to optimize SCWG processes for maximizing H_2 yield and carbon yield while minimizing CO_2 yield [49]. Research on air gasification of polyethylene waste highlights the potential of ML for predicting gasification performance with high accuracy [50].

A hybrid approach combining a simplified model with machine learning has been used to improve hydrogen formation prediction accuracy in microalgae biomass gasification [52]. Studies have employed ML for predicting gasification performance of polyethylene waste and ranking different test scenarios [54]. ANN models have been shown to outperform traditional methods in predicting coal and biomass gasification performance metrics [51].

ML has been used to develop relationships between catalyst properties, operating conditions, and syngas composition in thermo-catalytic biomass gasification for H_2 production [55]. AutoML techniques have been explored to identify the best ML model for predicting various outputs (syngas composition, LHV, char yield, tar yield) in fluidized-bed biomass gasification systems [56]. ML algorithms like gradient boosting have been used to predict hydrogen production from co-gasification of biomass and plastics based on feedstock properties and operating conditions [57]. XGBoost has been demonstrated as an effective tool for predicting syngas properties (H_2 concentration) in steam gasification of biomass [58].

2.4 *Industry 4.0 in Biomass Waste Management*

Industry 4.0 (I4.0) is transforming various sectors, including waste management, by merging digital technologies with physical processes. In the context of biomass waste management, I4.0 provides a more efficient and sustainable approach to handling waste through the integration of advanced technologies such as the Internet of Things (IoT), AI, ML, and big data analytics [59]. These technologies enable real-time monitoring, data-driven decision-making, and optimized process control, which collectively enhance the efficiency of waste management systems, reduce operational costs, and contribute to environmental sustainability.

2.4.1 Key Components of Industry 4.0 and Their Applications

IoT involves embedding sensors in waste bins to monitor fill levels, enabling optimized collection routes. Real-time data on waste composition can be gathered through smart bins, facilitating more effective sorting and recycling efforts [60]. This integration of IoT technology helps streamline waste management operations, making the process more efficient and responsive to real-time needs.

AI and ML play a crucial role in analyzing waste composition data to predict waste generation and optimize collection schedules. ML algorithms can automate waste sorting processes, separating recyclables from organic materials with greater accuracy [61]. Leveraging AI and ML enables waste management systems to become more intelligent and adaptive, leading to better resource utilization and reduced operational costs.

Big data analytics involve analyzing data collected from sensors and waste management systems to identify trends and inefficiencies. This data can optimize logistics, improve resource allocation, and predict maintenance needs [62]. By utilizing big data analytics, waste management companies can make data-driven decisions, enhancing the overall efficiency and effectiveness of their operations.

2.4.2 Benefits of Integrating Industry 4.0 in Biomass Waste Management

Integrating I4.0 technologies leads to improved efficiency by optimizing collection routes, streamlining sorting processes, and predicting waste generation, resulting in significant cost savings. These enhancements ensure a more efficient use of resources and reduce the operational costs associated with waste management [63, 64].

Real-time monitoring provided by sensors offers real-time data on waste levels and composition, enabling proactive management and preventing overflows or missed collections. This ensures that waste collection services are timely and responsive, reducing the likelihood of environmental pollution and health hazards [60, 65].

Optimized process control is achieved through AI and big data analytics, optimizing various processes within the waste management system, such as sorting, composting, and energy production from biomass. By optimizing these processes, waste management systems can enhance their performance, leading to increased productivity and sustainability [66, 67].

For instance, a study by Kurniawan et al. [61] explores how digitalization can improve waste management in Malang, Indonesia, drawing valuable lessons from Nanning, China's digital waste recycling model. Additionally, Mohammadi et al. [60] propose a holistic waste management system integrating IoT-equipped bins with route optimization algorithms. This approach optimizes collection routes, reduces costs, and minimizes environmental impact.

2.4.3 Challenges and Future Directions

There are several challenges to implementing I4.0 in biomass waste management. Data security and privacy are major concerns due to the extensive data collection involved. Robust cybersecurity measures are necessary to protect sensitive information about waste generation and composition. Infrastructure investment is another challenge, as implementing I4.0 technologies requires significant investment in sensors, communication networks, and data management systems, which can be a hurdle for cities and waste management companies with limited budgets. Additionally, the integration and standardization of different I4.0 technologies and data formats are crucial for seamless data integration and efficient information exchange. Public awareness and acceptance are also important, as educating the public about the benefits of I4.0 in waste management and addressing potential privacy concerns is essential for gaining public trust and acceptance.

Future directions for I4.0 in biomass waste management should focus on developing and deploying open-source solutions to make these technologies more accessible to smaller waste management companies and developing countries. Public–private partnerships can facilitate investment in I4.0 infrastructure and promote knowledge sharing. Integrating I4.0 technologies with circular economy principles can create a circular economy for biomass waste, maximizing resource recovery and minimizing waste disposal. Additionally, it is important to consider the environmental impact of I4.0 implementation itself, ensuring the use of energy-efficient technologies and responsible e-waste management practices.

3 Methodology

This section details the methodological approach employed to develop and evaluate machine learning models for optimizing biomass waste co-gasification. The methodology comprises six main stages: data generation, data preprocessing (including feature selection and data cleaning), data splitting, model training, model validation, and hyperparameters optimization, as illustrated in Fig. 1. Each stage plays a

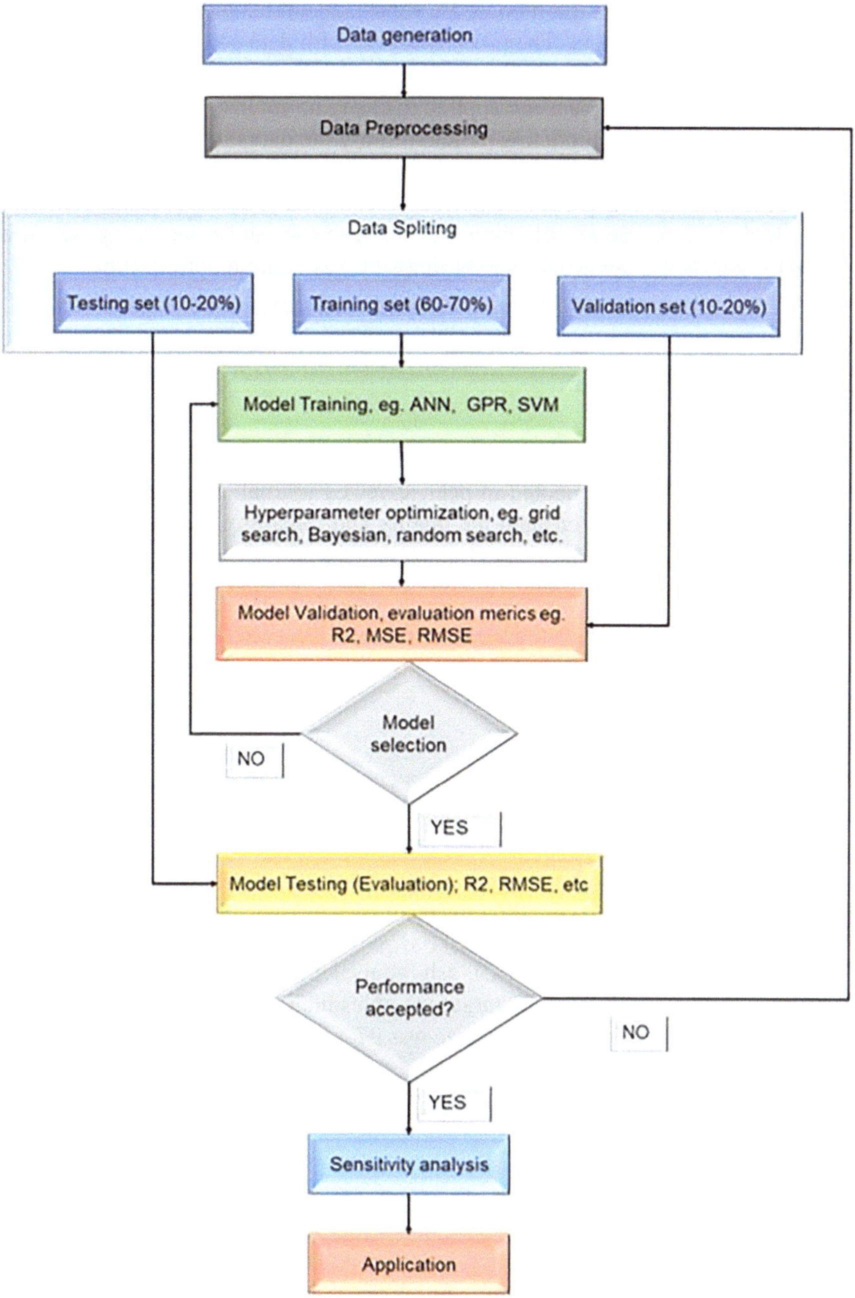

Fig. 1 Flowchart illustrating the machine learning model development process

vital role in ensuring that the models are robust, accurate, and capable of generalizing to new or unseen data. The process begins with data generation, where high-quality experimental data are sourced to ensure reliable model inputs. Subsequently, data preprocessing techniques are applied to clean, normalize, and encode the data, preparing it for effective modeling. After splitting the data into training, validation, and test sets, various machine learning algorithms are trained and optimized to achieve the best possible predictive performance. Finally, model evaluation metrics are applied to assess the models' accuracy and generalizability, ensuring their suitability for real-world applications in biomass co-gasification optimization.

3.1 Data Generation

Experimental data for this study were generated from biomass co-gasification experimental results documented in peer-reviewed journal articles from reputable publishers. This approach ensured the quality and reliability of the data. The variables considered for predictive modeling included a mix of numeric and categorical features relevant to the co-gasification process. Only data from high-quality sources were included to maintain the integrity of the dataset.

3.2 Feature Selection

The original dataset comprised 27 features. Feature selection was performed by removing features with more than 85% missing data, resulting in a refined dataset with 21 features and 4 target variables. The dataset included 7 categorical features and 14 numeric features. The numeric features consisted of particle size, airflow rate, carbon (C), hydrogen (H), nitrogen (N), sulfur (S), oxygen (O), moisture, volatile matter (VM), fixed carbon (FC), ash, equivalent ratio (ER), temperature, and higher heating value (HHV). The target variables included the syngas composition (CO, CO_2, CH_4, and H_2). The dataset size was 97, and the descriptive statistics of the numeric features and target variables are presented in Table 1.

3.3 Data Preprocessing

Data preprocessing is a crucial step in ensuring the accuracy and reliability of machine learning models. This study employed a multi-step approach to preprocessing, ensuring that the dataset was clean, normalized, and appropriately encoded. The general preprocessing steps is illustrated in Fig. 2.

Table 1 Descriptive statistic for numeric features and target variables

Variables	Number	Minimum	Maximum	Mean	Std. deviation
Features					
Particle size	92	0.40	7.50	3.2572	1.8130
Air flow rate	92	24.63	15,700.00	2289.5683	5314.2111
C	97	38.50	52.30	44.4866	1.9597
H	97	4.31	6.50	5.3420	0.7257
N	97	0.33	1.01	0.6078	0.1364
S	85	0.00	0.45	0.2328	0.0759
O	92	36.10	50.44	47.9342	3.3752
MC	97	2.64	12.20	5.4903	2.8314
VM	97	60.21	86.25	75.1898	7.6433
FC	97	8.45	26.08	17.7474	4.0746
ASH	97	1.10	19.70	4.4363	3.7671
HHV	91	15.61	20.38	17.9430	0.7543
T	92	700.00	900.00	795.4348	63.020
ER	41	0.19	0.37	0.3332	0.0621
Target					
CO	68	1.85	22.05	13.7028	5.0687
CO_2	17	6.01	15.03	10.2806	2.9224
CH_4	85	0.69	21.70	12.9293	5.2217
H_2	68	1.54	20.64	11.5975	3.4048

3.3.1 Cleaning of Missing Data for Numeric Data

Missing data in numeric features were addressed using linear interpolation, a method that estimates missing values based on the values of surrounding data points. This approach helps maintain the continuity of the data and avoids potential biases that can arise from simply removing missing values. The maximum gap to fill was set to infinity, indicating that all missing values were subject to interpolation regardless of the gap size. The missing value is estimated using the following linear interpolation formula in Eq. (1):

$$y = y_1 + \frac{(x - x_1)}{(x_2 - x_1)} \times (y_2 - y_1) \tag{1}$$

where x_1 and x_2 are the positions (indices or values) of the known data points surrounding the missing value, y_1 and y_2 are the known values at points x_1 and x_2, respectively, x is the position of the missing data point where the value y needs to be estimated, y is the missing value.

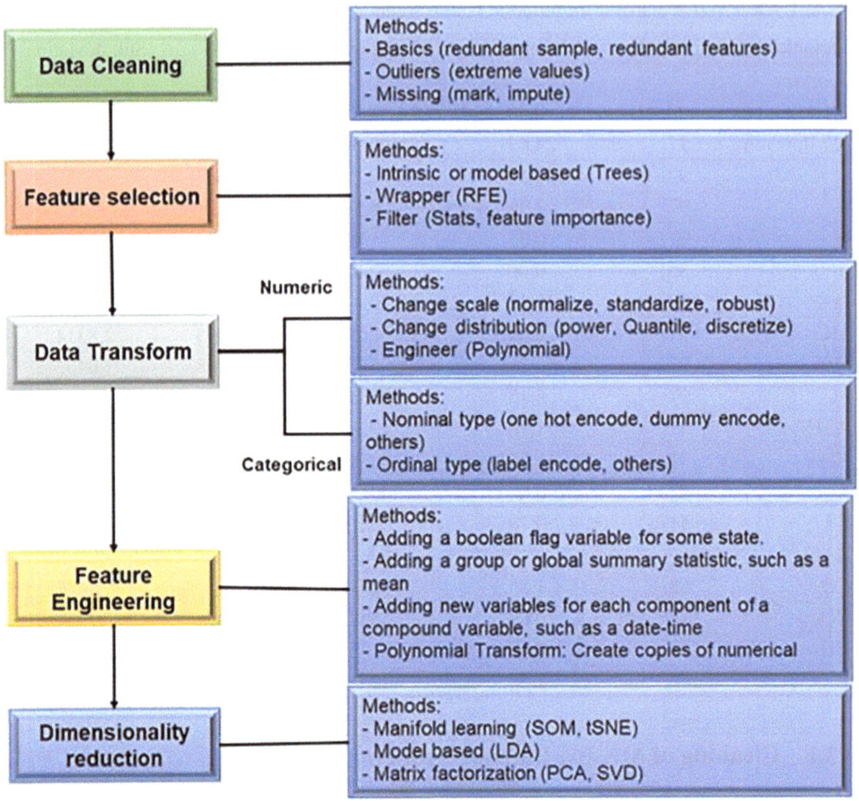

Fig. 2 General data preprocessing steps for machine learning

3.3.2 Cleaning of Outliers

Outliers in numeric features were identified and addressed using a moving median method. This method involves calculating the median value within a moving window across the data. Outliers were then filled using linear interpolation, similar to the approach for missing data. This step ensures that extreme values that could skew the model's predictions are managed effectively. The moving window type was centered, and the window length was set to 4.

3.3.3 Normalization

Numeric features were normalized using the range method, scaling values to a standard range between 0 and 1. This normalization helps to ensure that all features contribute equally to the model training process and prevents features with larger ranges from disproportionately influencing the model. The normalization is estimated using Eq. (2)

$$x' = \frac{x - \min(x)}{\max(x) - \min(x)} \tag{2}$$

where x' is the normalized value of the feature, x is the original value of the feature, min(x) is the minimum value of the feature in the dataset, and max(x) is the maximum value of the feature in the dataset.

3.3.4 Encoding for Categorical Data

Categorical features were encoded using label encoding, a method that converts categorical values into numeric labels. This transformation is crucial for machine learning models, which require numerical input to process categorical data effectively. The categorical features included in this study were blend type, feedstock shape, blend ratio, feeding type, gasifying agent, reactor type, and reactor scale. The blend type feature comprised various combinations of biomass materials used in the co-gasification process. The specific blends were coffee husk and sawdust, oil palm trunk and oil palm frond, wood coconut fiber and rice husk pellet, mesocarp fiber and palm kernel shell, coconut shell and oil palm frond, rice husk and bamboo dust, rice husk and sawdust, and bamboo dust and sawdust. These combinations represent the diverse sources of biomass feedstock utilized in different gasification experiments.

The feedstock shape feature captured the physical form of the biomass feedstock. The shapes considered were particle, fiber/pellet, and fiber/particle. Each shape type can influence the gasification process differently due to variations in surface area and density. The blend ratio feature indicated the proportion of each component in the biomass blend. The blend ratios included 0/100, 20/80, 25/75, 30/70, 50/50, 70/30, 75/25, 80/20, and 100/0. These ratios are critical for understanding how different proportions of biomass materials affect the syngas composition and overall gasification efficiency.

The feeding type feature described the method of introducing biomass into the gasifier, which could be either continuous or batch. Continuous feeding involves a steady supply of biomass, whereas batch feeding involves adding biomass in discrete amounts. This distinction is important as it can impact the consistency and control of the gasification process. The gasifying agent feature captured the type of agent used to facilitate the gasification reaction. The agents considered were air and a combination of air and steam. The choice of gasifying agent affects the chemical reactions within the gasifier and, consequently, the composition of the produced syngas.

The reactor type feature identified the design of the gasification reactor. The types included bubbling fluidized bed, downdraft, and circulating fluidized-bed reactors. Each reactor type has unique operational characteristics that influence the gasification process and syngas production. Lastly, the reactor scale feature indicated the size of the gasification reactor, which could be lab-scale, bench, or pilot-scale. The scale of the reactor is essential for translating experimental results to industrial applications and understanding the scalability of the gasification process. By encoding these categorical features, the study ensured that the machine learning models could interpret and utilize the diverse range of experimental conditions represented in the dataset. This encoding process was a critical step in preparing the data for effective predictive modeling.

3.4 Model Training, Validation, and Hyperparameter Optimization

Three machine learning algorithms were employed in this study: ANN, GPR, and GBR. These algorithms were selected for their proven efficacy in handling complex, non-linear relationships within datasets, making them suitable for predicting syngas composition from biomass co-gasification.

A five-fold cross-validation technique was used to evaluate the models, ensuring robust performance metrics. Additionally, 10% of the total data was set aside as a test dataset for final model evaluation.

Bayesian optimization was applied to all models for hyperparameter tuning. Bayesian optimization was chosen over other methods such as grid search and random search due to its efficiency and effectiveness in finding optimal hyperparameters. Unlike grid search, which exhaustively searches over a specified parameter grid, and random search, which randomly samples hyperparameters, Bayesian optimization builds a probabilistic model of the objective function and uses it to select the most promising hyperparameters to evaluate next. This approach significantly reduces the number of evaluations required to find the optimal hyperparameters, making it more suitable for complex models and large datasets.

Model development was performed using MATLAB version 2022b, leveraging the Statistical and Machine Learning Toolbox for implementing various machine learning techniques. The regression learner application was utilized to streamline the process of building and evaluating the regression models. This tool provided an interactive environment for model selection, training, and validation, allowing for efficient comparison of multiple algorithms.

3.5 Performance Evaluation Metrics

A wide range of performance metrics (uncertainty indices or statistical measures) are available for evaluating the accuracy of a model. The sheer number of these metrics can create confusion regarding their appropriate use, making it a common practice to present multiple metrics and allow the reader to interpret them accordingly [68–70]. However, the selection of these metrics is typically guided by their proven effectiveness in related studies for assessing the suitability of a model for prediction [71]. Thus, employing multiple metrics to evaluate prediction accuracy is highly advantageous. In this study, three key metrics were used to evaluate the performance of the models. These include coefficient of determination (R^2), Root Mean Squared Error (RMSE), and Mean Squared Error (MSE).

3.5.1 The Coefficient of Determination or R-Squared (R^2)

R^2 is the metric used to measure the proportion of the variance in the dependent variable that is predictable from the independent variables in the regression model. It measures the correlation between outputs and targets. R^2 shows how well the model's predictions match the actual data. It value ranges from 0 to 1, and a higher R^2 value (closer to 1) indicates a better fit of the model to the data, meaning that the model's predictions are closer to the actual values. The R^2 (Eq. 3) is commonly used to assess the goodness-of-fit of regression models [72, 73].

$$R^2 = 1 - \left[\frac{\sum_{i=1}^{m} \left(y_i - \hat{y}_i\right)^2}{\sum_{i=1}^{m} \left(y_i - \overline{y}\right)^2} \right] \tag{3}$$

where R^2 is the coefficient of determination, y_i is the actual values of the dependent variable, $\hat{y}_i$ is the predicted values from the model, $\overline{y}$ is the mean of the actual values (y_i), m is the number of data points, i is an index used to denote each individual data point in a dataset (ranges from *1 to m*).

3.5.2 Mean Squared Error (MSE)

MSE is a common metric used to evaluate the performance of regression models. It is the average of the squared differences between the predicted and actual values. A lower MSE indicates better model performance, and a zero value means no error. However, MSE gives more weight to larger errors due to the squaring of differences. It can be used if there are outliers that need to be detected [68]. It is calculated using Eq. (4)

$$\text{MSE} = \frac{\sum_{i=1}^{m} \left(y_i - \hat{y}_i\right)^2}{m} \tag{4}$$

where y_i is the actual values of the dependent variable, $\hat{y}_i$ is the predicted values from the model, m is the number of data points.

3.5.3 Root Mean Squared Error (RMSE)

RMSE is the square root of the average of the squared differences between the predicted and actual values. It provides a measure of the average magnitude of the errors in a set of predictions, with a lower RMSE indicating better model performance. RMSE is sensitive to outliers since it involves squaring the differences, which amplifies larger errors more than smaller ones [72, 74, 75]. It is estimated using Eq. (5)

$$\text{RMSE} = \sqrt{\frac{\sum_{i=1}^{m}\left(y_i - \hat{y}_i\right)^2}{m}} \tag{5}$$

where y_i is the actual value of the dependent variable, $\hat{y}_i$is the predicted values from the model, m is the number of data points.

4 Results and Discussion

This section presents the results of the machine learning models used for predicting syngas composition in biomass co-gasification and discusses their performance and implications for Industry 4.0. The effectiveness of three different models including GPR, GBR, and ANN was evaluated using various performance metrics. The analysis covers both the validation and test datasets to provide a comprehensive understanding of each model's predictive capabilities. Additionally, a comparative analysis of the models' accuracy and reliability is conducted, highlighting the strengths and weaknesses of each approach. The findings are further contextualized within the framework of Industry 4.0, exploring the potential applications, benefits, and challenges associated with integrating these machine learning models into co-gasification processes. The implications for future research and practical deployment in industrial settings are also discussed.

4.1 Model Performance

The performance metrics of three different machine learning models including GPR, GBR, and ANN were evaluated for predicting syngas composition using validation and test datasets. For the validation dataset, the metrics used were MSE and R^2, whereas for the test dataset, RMSE and R^2 were used.

In the validation dataset (Table 2), the ANN model demonstrated superior performance in predicting CO with the lowest MSE of 0.0165 and the highest R^2 of 0.8024, surpassing GPR and GBR. For CO_2 prediction, GPR showed the best results with an MSE of 0.0010 and an R^2 of 0.9892, closely followed by GBR and ANN. When

Table 2 Model performance on syngas composition using validation dataset

	GPR		GBR		ANN	
Gas composition	MSE	R^2	MSE	R^2	MSE	R^2
CO	0.0174	0.7927	0.0190	0.7783	0.0165	0.8024
CO_2	0.0010	0.9892	0.0011	0.9884	0.0048	0.9482
H_2	0.0102	0.8661	0.0097	0.8637	0.0104	0.8636
CH_4	0.0365	0.4500	0.0272	0.6284	0.0357	0.4624
Average	0.0651	0.7745	0.0143	0.8147	0.0168	0.7691

Table 3 Model performance on syngas composition prediction using test dataset

	GPR		GBR		ANN	
Gas composition	RMSE	R^2	RMSE	R^2	RMSE	R^2
CO	0.1199	0.8354	0.1002	0.9021	0.1149	0.8488
CO_2	0.0186	0.9958	0.0339	0.9873	0.0334	0.9863
H_2	0.0653	0.9472	0.0697	0.9246	0.0696	0.9399
CH_4	0.2225	0.4295	0.1924	0.5732	0.2258	0.4120
Average	0.0998	0.8020	0.099	0.8468	0.1109	0.7968

predicting H_2, both GPR and GBR performed similarly, with GBR having a slightly better MSE of 0.0097 and comparable R^2 values around 0.866. For CH_4 prediction, GBR outperformed the other models, achieving the highest R^2 of 0.6284 and the lowest MSE of 0.0272.

For the test dataset (Table 3), the GBR model demonstrated the best performance in predicting CO, with the lowest RMSE of 0.1002 and the highest R^2 of 0.9021. In CO_2 prediction, GPR emerged as the best performer with an RMSE of 0.0186 and an R^2 of 0.9958. For H_2 prediction, GPR again outperformed the other models with an RMSE of 0.0653 and an R^2 of 0.9472. However, for CH_4 prediction, GBR showed the best performance with the highest R^2 of 0.5732 and a lower RMSE of 0.1924 compared to other models.

4.2 *Comparative Analysis*

The performance of the machine learning models in predicting both CO and CO_2 levels was evaluated by comparing the predicted responses from GPR, GBR, and ANN models against the true responses. This comparison is illustrated in Figs. 3 and 4, which show the predicted values of CO and CO_2 levels plotted against their corresponding true values. For both gases, the accuracy of the predictions was assessed based on how closely the predicted responses matched the true responses. As shown in the figures, the GPR, GBR, and ANN models demonstrate their capability to predict CO and CO_2 levels, with varying levels of accuracy.

In comparing the efficacy of the three models, GBR consistently showed strong performance, particularly in predicting CO and CH_4 (Fig. 5), making it a

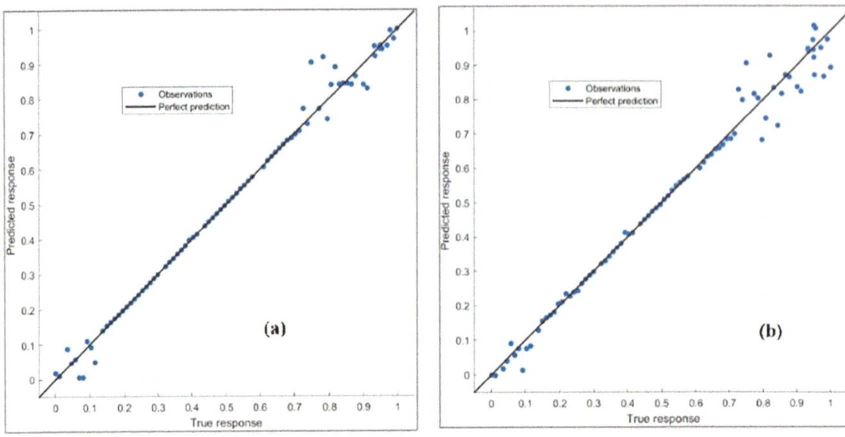

Fig. 3 Comparison of predicted and true responses for CO_2 prediction using (**a**) GPR and (**b**) ANN

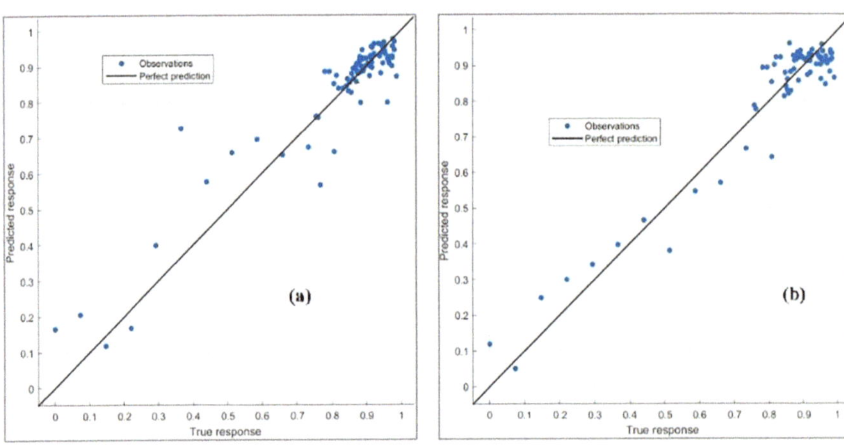

Fig. 4 Comparison of predicted and true responses for CO prediction using (**a**) GBR and (**b**) ANN

well-balanced model for comprehensive syngas composition prediction. GPR excelled in predicting CO_2 and H_2, demonstrating superior predictive power and lower error rates. While ANN showed strong performance in certain areas, such as CO prediction in the validation dataset, it was generally outperformed by GBR and GPR in the test dataset (Fig. 6). The choice of model should be based on the specific gas composition of interest, with GBR being a robust option for a balanced approach.

The findings of this study demonstrate that machine learning models can effectively predict syngas composition, with GBR and GPR showing particularly strong performance across various gas compositions. These results align with existing literature, which highlights the potential of machine learning in enhancing the accuracy and reliability of predictive models in energy and environmental applications. The superior performance of GBR, especially in predicting CO and CH_4,

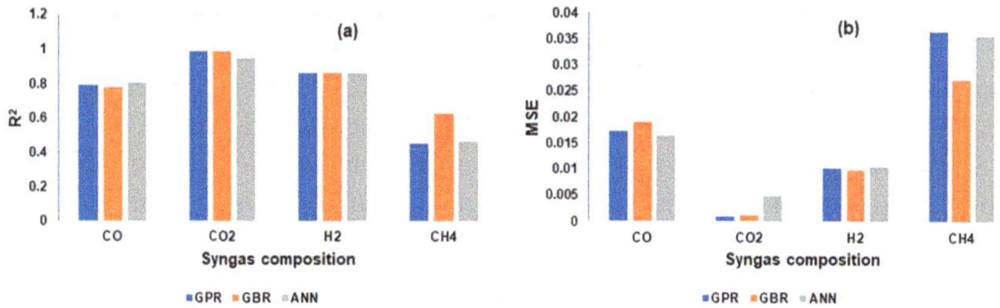

Fig. 5 Comparison of the prediction accuracy for the different models based on validation using the performance metrics: (**a**) R^2, (**b**) MSE

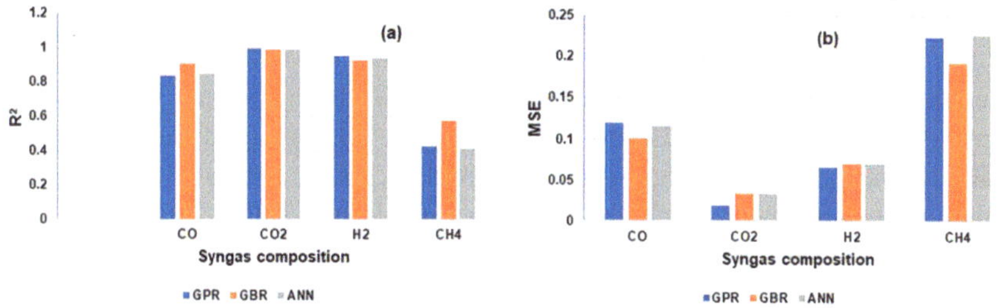

Fig. 6 Comparison of the prediction accuracy for the different models based on test data using the performance metrics: (**a**) R^2, (**b**) RMSE

underscores its potential as a valuable tool in the optimization of syngas production processes.

A comparative analysis of our study's findings with existing literature reveals several key insights. For instance, Wen et al. [76] reported R^2 values ranging from 0.80 to 0.89 for ANN and 0.81 to 0.93 for GBR in predicting gas composition from rice husks. This range is comparable to our study's results, where GBR outperformed ANN in predicting syngas components. Similarly, Ascher, Wang et al. [77] and Ascher, Sloan et al. [78] found that GBR outperformed other models, including Extreme Gradient Boosting (XGBoost), AdaBoost, RF, and ANN, across various gasifier types and feedstocks. Their findings are consistent with our results, where GBR showed high predictive accuracy for multiple target variables.

Li et al. [25] highlighted GBR's strong performance with test R^2 values between 0.82 and 0.96 for predicting gas composition from food waste, sludge, and mixed biomass waste. Our study's GBR model demonstrated similar robust performance, further validating GBR's effectiveness in diverse gasification contexts. Haq et al.

[79] found that GPR had the highest accuracy with an R^2 value of 0.997 for predicting H_2 yield from sewage sludge, while SVM had the lowest accuracy ($R^2 = 0.761$). These results underscore the high accuracy of GPR in specific prediction tasks, paralleling our findings where GPR excelled in predicting CO_2 and H_2.

Sison et al. [48] reported that GBR achieved the highest R^2 score of 0.9453 for both H_2 and char yield in chemical looping gasification of various biomass feedstock. This supports our conclusion that GBR is a highly reliable model for predicting multiple outputs in gasification processes. Similarly, Pandey et al. [80] and Yang et al. [81] both reported that GBR had the highest prediction accuracy for various outputs, with R^2 scores often exceeding 0.84 and 0.926, respectively. These consistent findings across different studies highlight GBR's robust performance in gasification modeling.

4.3 Implications for Industry 4.0

Integrating machine learning models into co-gasification systems has significant implications for Industry 4.0 implementation. By leveraging advanced predictive analytics, industries can achieve improved process control, enhanced efficiency, and optimized resource utilization. The adoption of GBR and GPR models can facilitate real-time monitoring and adjustment of gasification processes, leading to more consistent syngas quality and reduced operational costs. This integration supports the broader goals of Industry 4.0, including increased automation, data-driven decision-making, and the development of smart manufacturing ecosystems.

4.4 Challenges and Limitations

Despite the promising results, several challenges and limitations were encountered during this study. One notable issue was the variability in model performance across different gas compositions, highlighting the need for tailored approaches for each specific prediction task. The models' performance on the test dataset, particularly for CH_4 prediction, was lower than on the validation dataset, suggesting potential issues with generalizability. This indicates a need for further research to enhance model robustness and explore the impact of different data preprocessing techniques and model architectures.

The quality and quantity of data significantly influenced model performance. Despite rigorous preprocessing, data inconsistencies and limited sample sizes may have impacted the accuracy of the models. The study primarily relied on data from the literature, which may not fully capture the variability and complexity of real-world gasification processes. To improve model generalizability, future research should focus on collecting more extensive and diverse datasets. Additionally, the computational complexity of advanced models like GBR and GPR poses challenges for real-time application in industrial settings. Implementing these models requires substantial computational resources and expertise in machine learning, which may not be readily available in all industries. Addressing these challenges is crucial for the widespread adoption of machine learning models in gasification and other industrial processes.

This study contributes to the growing body of knowledge on the application of machine learning in gasification, demonstrating the effectiveness of GBR and GPR models in predicting syngas composition. However, ongoing research and development are essential to overcome the identified challenges and limitations, ensuring the reliable and efficient application of these technologies in real-world gasification systems. The study demonstrates the efficacy of machine learning models, particularly GBR and GPR, in predicting syngas composition, with significant implications for Industry 4.0. However, addressing the challenges of data quality, computational complexity, and model generalizability is essential for the successful deployment of these models in industrial contexts. Future research should continue to explore the potential of these models, focusing on overcoming current limitations and enhancing their practical applicability in Industry 4.0 environments.

5 Conclusion and Recommendations

This study evaluated the performance of the machine learning models in predicting syngas composition. The results indicate that GBR consistently showed strong performance, particularly in predicting CO and CH_4, making it the most balanced model for comprehensive syngas composition prediction. GPR excelled in predicting CO_2 and H_2, demonstrating superior predictive power and lower error rates. While ANN showed strong performance in certain areas, such as CO prediction in the validation dataset, it was generally outperformed by GBR and GPR in the test dataset. These findings highlight the potential of using advanced machine learning models to enhance the accuracy and reliability of syngas composition predictions, with significant implications for optimizing gasification processes and supporting Industry 4.0 implementation.

Future research should focus on enhancing the robustness and generalizability of the machine learning models by exploring various data preprocessing techniques and model architectures. Investigating the integration of hybrid models, which combine the strengths of different machine learning approaches, could further improve predictive accuracy. Additionally, expanding the dataset to include a broader range of operational conditions and syngas compositions will help to validate the models' performance in diverse scenarios. Research should also explore the real-time application of these models in industrial settings, examining their impact on process optimization and operational efficiency. Finally, studies on the integration of machine learning models with other Industry 4.0 technologies, such as digital twins and the Internet of Things (IoT), could provide valuable insights into developing smart, automated, and highly efficient gasification systems.

Acknowledgments The authors acknowledge the financial support from African Center of Excellence II in Phytochemical, Textile and Renewable Energy (ACE II-PTRE), and Strengthening Mobility and Promoting Regional Integration of Engineering Education in Africa (SPREE).

Disclosure of Interests The authors have no competing interests to declare that are relevant to the content of this article.

References

1. Bastos, T., Teixeira, L.C., Nunes, L.J.R.: Forest 4.0: technologies and digitalization to create the residual biomass supply chain of the future. J. Clean. Prod. **467**, 143041 (2024). https://doi.org/10.1016/j.jclepro.2024.143041
2. Javaid, M., Haleem, A., Singh, R.P., Suman, R., Gonzalez, E.S.: Understanding the adoption of Industry 4.0 technologies in improving environmental sustainability. Sustain. Oper. Comput. **3**, 203–217 (2022). https://doi.org/10.1016/j.susoc.2022.01.008
3. Bongomin, O., Nganyi, E.O., Abswaidi, M.R., Hitiyise, E., Tumusiime, G.: Sustainable and dynamic competitiveness towards technological leadership of Industry 4.0: implications for East African Community. J. Eng. **2020**, 1–22 (2020). https://doi.org/10.1155/2020/8545281
4. Bongomin, O., Yemane, A., Kembabazi, B., Malanda, C., Chikonkolo Mwape, M., Sheron Mpofu, N., Tigalana, D.: Industry 4.0 disruption and its neologisms in major industrial sectors: a state of the art. J. Eng. **2020**, 1–45 (2020). https://doi.org/10.1155/2020/8090521
5. Bongomin, O., Gilibrays Ocen, G., Oyondi Nganyi, E., Musinguzi, A., Omara, T.: Exponential disruptive technologies and the required skills of Industry 4.0. J. Eng. **2020**, 1–17 (2020). https://doi.org/10.1155/2020/4280156
6. Sánchez-García, E., Martínez-Falcó, J., Marco-Lajara, B., Manresa-Marhuenda, E.: Revolutionizing the circular economy through new technologies: a new era of sustainable progress. Environ. Technol. Innov. **33**, 103509 (2024). https://doi.org/10.1016/j.eti.2023.103509
7. Bongomin, O., Nzila, C., Mwasiagi, J.I., Maube, O.: Exploring insights in biomass and waste gasification via ensemble machine learning models and interpretability techniques. Int. J. Energy Res. **2024** (2024). https://doi.org/10.1155/2024/6087208
8. Cai, J., Zhu, L., Yang, J., Guo, M., Fang, M., Yao, S.: Synergistic co-steam gasification of biomass and refuse-derived fuel: a path to enhanced gasification performance. Environ. Technol. Innov. **36**, 103745 (2024). https://doi.org/10.1016/j.eti.2024.103745
9. Abioye, K.J., Harun, N.Y., Sufian, S., Yusuf, M., Jagaba, A.H., Waqas, S., Ayodele, B.V., Kamyab, H., Alam, M., Gupta, M., Gill, H.S., Rezania, S., Chelliapan, S., Kang, K.: Optimization of syngas production from co-gasification of palm oil decanter cake and alum sludge: an RSM approach with char characterization. Environ. Res. **246**, 118027 (2024). https://doi.org/10.1016/j.envres.2023.118027
10. Rashwan, A.K., Bai, H., Osman, A.I., Eltohamy, K.M., Chen, Z., Younis, H.A., Al-Fatesh, A., Rooney, D.W., Yap, P.S.: Recycling food and agriculture by-products to mitigate climate change: a review. Environ. Chem. Lett. **21**, 3351–3375 (2023). https://doi.org/10.1007/s10311-023-01639-6
11. Anvari, S., Aguado, R., Jurado, F., Fendri, M., Zaier, H., Larbi, A., Vera, D.: Analysis of agricultural waste/byproduct biomass potential for bioenergy: the case of Tunisia. Energy Sustain. Dev. **78**, 101367 (2024). https://doi.org/10.1016/j.esd.2023.101367
12. Peng, X., Jiang, Y., Chen, Z., Osman, A.I., Farghali, M., Rooney, D.W., Yap, P.S.: Recycling municipal, agricultural and industrial waste into energy, fertilizers, food and construction materials, and economic feasibility: a review. Environ. Chem. Lett. **21**, 765 (2023)

13. Hajam, Y.A., Kumar, R., Kumar, A.: Environmental waste management strategies and vermi transformation for sustainable development. Environ. Challenges. **13**, 100747 (2023). https://doi.org/10.1016/j.envc.2023.100747
14. Nguyen, V.G., Sharma, P., Ağbulut, Ü., Le, H.S., Truong, T.H., Dzida, M., Tran, M.H., Le, H.C., Tran, V.D.: Machine learning for the management of biochar yield and properties of biomass sources for sustainable energy. Biofuels Bioprod. Biorefin. **18**, 567–593 (2024). https://doi.org/10.1002/bbb.2596
15. Adeleke, A.A., Adedigba, A., Adeshina, S.A., Ikubanni, P.P., Lawal, M.S., Olosho, A.I., Yakubu, H.S., Ogedengbe, T.S., Nzerem, P., Okolie, J.A.: Comparative studies of machine learning models for predicting higher heating values of biomass. Digit. Chem. Eng. **12**, 100159 (2024). https://doi.org/10.1016/j.dche.2024.100159
16. Zhang, P., Zhang, T., Zhang, J., Liu, H., Chicaiza-Ortiz, C., Lee, J.T.E., He, Y., Dai, Y., Tong, Y.W.: A machine learning assisted prediction of potential biochar and its applications in anaerobic digestion for valuable chemicals and energy recovery from organic waste. Carbon Neutrality. **3** (2024). https://doi.org/10.1007/s43979-023-00078-0
17. Tewari, K., Balyan, S., Jiang, C., Robinson, B., Bhattacharyya, D., Hu, J.: Unlocking syngas synthesis from the catalytic gasification of lignocellulose pinewood: catalytic and pressure insights. ACS Sustain. Chem. Eng. **12**, 4718–4730 (2024). https://doi.org/10.1021/acssuschemeng.4c00320
18. Gao, Y., Wang, M., Raheem, A., Wang, F., Wei, J., Xu, D., Song, X., Bao, W., Huang, A., Zhang, S., Zhang, H.: Syngas production from biomass gasification: influences of feedstock properties, reactor type, and reaction parameters. ACS Omega. **8**, 31620–31631 (2023). https://doi.org/10.1021/acsomega.3c03050
19. Williams, J.M., Bourtsalas, A.C.: Assessment of co-gasification methods for hydrogen production from biomass and plastic wastes. Energies. **16** (2023). https://doi.org/10.3390/en16227548
20. Bongomin, O., Nziu, P.: A critical review on the development and utilization of energy systems in Uganda. Sci. World J. **2022**, 1–25 (2022). https://doi.org/10.1155/2022/2599467
21. Kumar, P., Subbarao, P.M.V., Kala, L.D., Vijay, V.K.: Real-time performance assessment of open-top downdraft biomass gasifier system. Clean. Eng. Technol. **7**, 100448 (2022). https://doi.org/10.1016/j.clet.2022.100448
22. Singh, D.K., Raj, R., Tirkey, J.V., Jena, P., Parthasarathy, P., Mckay, G., Al-Ansari, T.: Progress and utilization of biomass gasification for decentralized energy generation: an outlook & critical review. Environ. Technol. Rev. **12**, 1–36 (2023). https://doi.org/10.1080/21622515.2023.2242014
23. Guo, X., Xu, Z., Zheng, X., Jin, X., Cai, J.: Understanding pyrolysis mechanisms of corn and cotton stalks via kinetics and thermodynamics. J. Anal. Appl. Pyrolysis. **164**, 105521 (2022). https://doi.org/10.1016/j.jaap.2022.105521
24. Das, N., Jena, P.K., Padhi, D., Kumar Mohanty, M., Sahoo, G.: A comprehensive review of characterization, pretreatment and its applications on different lignocellulosic biomass for bioethanol production. Biomass Convers. Biorefinery. **13**, 1503–1527 (2023). https://doi.org/10.1007/s13399-021-01294-3
25. Li, J., Li, L., Tong, Y.W., Wang, X.: Understanding and optimizing the gasification of biomass waste with machine learning. Green Chem. Eng. **4**, 123–133 (2023). https://doi.org/10.1016/j.gce.2022.05.006
26. Yao, S., Zhang, Y., Xia, J., Xie, T., Zhang, Z., Li, H., Hu, J.: Cascade utilization of energy in high temperature syngas to reduce energy consumption in biomass gasification processes. Case Stud. Therm. Eng. **52**, 103680 (2023). https://doi.org/10.1016/j.csite.2023.103680
27. De Padova, A., Giglio, E., Santarelli, M.: Electrolysis-boosted substitute natural gas from biomass: kinetic modeling of fluidized bed gasification and system integration. Renew. Energy. **223**, 120071 (2024). https://doi.org/10.1016/j.renene.2024.120071
28. Tomasek, S., Bárkányi, Á., Egedy, A., Miskolczi, N.: Model-based determination of optimal operating parameters for different solid waste gasification. Chem. Eng. J. Adv. **17**, 100586 (2024). https://doi.org/10.1016/j.ceja.2024.100586

29. Zhu, F., Shen, L., Xu, P., Yuan, H., Hu, M., Qi, J., Chen, Y.: Numerical simulation of an improved updraft biomass gasifier based on Aspen Plus. Int. J. Environ. Res. Public Health. **19** (2022). https://doi.org/10.3390/ijerph192417089
30. Jamil, F., Inayat, A., Hussain, M., Akhter, P., Abideen, Z., Ghenai, C., Shanableh, A., Abdellatief, T.M.M.: Valorization of waste biomass to biofuels for power production and transportation in optimized way: a comprehensive review. Adv. Energy Sustain. Res. **5**, 2400104 (2024). https://doi.org/10.1002/aesr.202400104
31. Joshi, N.C., Sinha, S., Bhatnagar, P., Nath, Y., Negi, B., Kumar, V., Gururani, P.: A concise review on waste biomass valorization through thermochemical conversion. Curr. Res. Microb. Sci. **6**, 100237 (2024). https://doi.org/10.1016/j.crmicr.2024.100237
32. Srivastava, R.K., Shetti, N.P., Reddy, K.R., Nadagouda, M.N., Badawi, M., Bonilla-Petriciolet, A., Aminabhavi, T.M.: Valorization of biowastes for clean energy production, environmental depollution and soil fertility. J. Environ. Manag. **332**, 117410 (2023). https://doi.org/10.1016/j.jenvman.2023.117410
33. Begum, Y.A., Kumari, S., Jain, S.K., Garg, M.C.: A review on waste biomass-to-energy: integrated thermochemical and biochemical conversion for resource recovery. Environ. Sci. Adv. **3**, 1197 (2024). https://doi.org/10.1039/d4va00109e
34. Blasi, A., Verardi, A., Lopresto, C.G., Siciliano, S., Sangiorgio, P.: Lignocellulosic agricultural waste valorization to obtain valuable products: an overview. Recycling. **8**, 1–46 (2023). https://doi.org/10.3390/recycling8040061
35. Chojnacka, K.: Valorization of biorefinery residues for sustainable fertilizer production: a comprehensive review. Biomass Convers. Biorefinery. **13**, 14359–14388 (2023). https://doi.org/10.1007/s13399-023-04639-2
36. de Dieu Marcel Ufitikirezi, J., Filip, M., Ghorbani, M., Zoubek, T., Olšan, P., Bumbálek, R., Strob, M., Bartoš, P., Umurungi, S.N., Murindangabo, Y.T., Heřmánek, A., Tupý, O., Havelka, Z., Stehlík, R., Černý, P., Smutný, L.: Agricultural waste valorization: exploring environmentally friendly approaches to bioenergy conversion. Sustainability (Switzerland). **16**, 1–24 (2024). https://doi.org/10.3390/su16093617
37. Yang, H., Nurdiawati, A., Gond, R., Chen, S., Wang, S., Tang, B., Jin, Y., Zaini, I.N., Shi, Z., Wang, W., Martin, A., Younesi, R., Sandström, L., Jönsson, P.G., Yang, W., Han, T.: Carbon-negative valorization of biomass waste into affordable green hydrogen and battery anodes. Int. J. Hydrog. Energy. **49**, 459–471 (2024). https://doi.org/10.1016/j.ijhydene.2023.09.096
38. Omol, D.K., Acaye, O., Okot, D.F., Bongomin, O.: Production of fuel oil from municipal plastic wastes using thermal and catalytic pyrolysis. J. Energy Res. Rev. **4**, 1–8 (2020). https://doi.org/10.9734/jenrr/2020/v4i230120
39. Lassi, U., Sarkar, A.: Chapter 1: Introduction to waste biomass processing and valorization. In: Processing of Biomass Waste. Elsevier (2024)
40. Ayub, Y., Zhou, J., Shen, W., Ren, J.: Innovative valorization of biomass waste through integration of pyrolysis and gasification: process design, optimization, and multi-scenario sustainability analysis. Energy. **282**, 128417 (2023). https://doi.org/10.1016/j.energy.2023.128417
41. Zhou, J., Ayub, Y., Shi, T., Ren, J., He, C.: Sustainable co-valorization of medical waste and biomass waste: innovative process design, optimization and assessment. Energy. **288**, 129803 (2024). https://doi.org/10.1016/j.energy.2023.129803
42. Tatla, H.K., Ismail, S., Khan, M.A., Dhar, B.R., Gupta, R.: Coupling hydrothermal liquefaction and anaerobic digestion for waste biomass valorization: a review in context of circular economy. Chemosphere. **361**, 142419 (2024). https://doi.org/10.1016/j.chemosphere.2024.142419
43. Sharma, S., Asolekar, S.R., Thakur, V.K., Asokan, P.: Unlocking the potential: a paradigm-shifting approach for valorizing lignocellulosic waste biomass of constructed wetland enabling environmental and societal sustainability. Ind. Crop. Prod. **207**, 117709 (2024). https://doi.org/10.1016/j.indcrop.2023.117709
44. Dessie, W., Luo, X., He, F., Liao, Y., Duns, G.J., Qin, Z.: Lignin valorization: a crucial step towards full utilization of biomass, zero waste and circular bioeconomy. Biocatal. Agric. Biotechnol. **51**, 102777 (2023). https://doi.org/10.1016/j.bcab.2023.102777

45. de Castro, S.C., Stanisic, D., Tasic, L.: Sequential extraction of hesperidin, pectin, lignin, and cellulose from orange peels: towards valorization of agro-waste. Biofuels Bioprod. Biorefin. **18**, 804–817 (2024). https://doi.org/10.1002/bbb.2606
46. Umer, M., Ahmad, J., Mukhtar, H.: Innovative valorization of biomass waste-derived sodium silicate for geopolymer concrete synthesis: sustainability assessment and circular economy potential. J. Clean. Prod. **452**, 142181 (2024). https://doi.org/10.1016/j.jclepro.2024.142181
47. Liu, Y., Ali Siyal, A., Zhou, C., Liu, C., Fu, J., Zhang, Y., Yao, B., Chao, L., Yun, H., Dai, J., Bi, X.: Microwave co-pyrolysis of industrial sludge and waste biomass: product valorization and synergistic mechanisms. Chem. Eng. J. **485**, 150032 (2024). https://doi.org/10.1016/j.cej.2024.150032
48. Sison, A.E., Etchieson, S.A., Güleç, F., Epelle, E.I., Okolie, J.A.: Process modelling integrated with interpretable machine learning for predicting hydrogen and char yield during chemical looping gasification. J. Clean. Prod. **414**, 137579 (2023). https://doi.org/10.1016/j.jclepro.2023.137579
49. Wang, C., Hu, C., Zheng, Y., Jin, H., Wu, Z.: Predictive control of reactor network model using machine learning for hydrogen-rich gas and biochar poly-generation by biomass waste gasification in supercritical water. Energy. **282**, 128441 (2023). https://doi.org/10.1016/j.energy.2023.128441
50. Hasanzadeh, R., Azdast, T.: Machine learning utilization on air gasification of polyethylene terephthalate waste. Waste Manag. Bull. **2**, 75–82 (2024). https://doi.org/10.1016/j.wmb.2023.12.011
51. Yang, Q., Zhang, J., Zhou, J., Zhao, L., Zhang, D.: A hybrid data-driven machine learning framework for predicting the performance of coal and biomass gasification processes. Fuel. **346**, 128338 (2023). https://doi.org/10.1016/j.fuel.2023.128338
52. Santos, J.M., Zelioli, Í.A.M., Guimarães F, E.É.X., Freitas, A.C.D., Mariano, A.P.: Supercritical water gasification thermodynamic study and hybrid modeling of machine learning with the ideal gas model: application to gasification of microalgae biomass. Energy. **291**, 130287 (2024). https://doi.org/10.1016/j.energy.2024.130287
53. Gil, M.V., Jablonka, K.M., Garcia, S., Pevida, C., Smit, B.: Biomass to energy: a machine learning model for optimum gasification pathways. Digit. Discov. **2**, 929–940 (2023). https://doi.org/10.1039/d3dd00079f
54. Gharibi, A., Babazadeh, R., Hasanzadeh, R.: Machine learning and multi-criteria decision analysis for polyethylene air-gasification considering energy and environmental aspects. Process Saf. Environ. Prot. **183**, 46–58 (2024). https://doi.org/10.1016/j.psep.2023.12.069
55. Karka, S., Kamesh, R.: Data analysis and machine learning aided integrated catalyst activity and process modelling for selective H2 production from biomass gasification. Biomass Bioenergy. **187**, 107291 (2024). https://doi.org/10.1016/j.biombioe.2024.107291
56. Kim, J.Y., Shin, U.H., Kim, K.: Predicting biomass composition and operating conditions in fluidized bed biomass gasifiers: an automated machine learning approach combined with cooperative game theory. Energy. **280**, 128138 (2023). https://doi.org/10.1016/j.energy.2023.128138
57. Devasahayam, S., Albijanic, B.: Predicting hydrogen production from co-gasification of biomass and plastics using tree based machine learning algorithms. Renew. Energy. **222**, 119883 (2024). https://doi.org/10.1016/j.renene.2023.119883
58. Xue, P., Chen, T., Huang, X., Hu, Q., Hu, J., Zhang, H., Yang, H., Chen, H.: Prediction of syngas properties of biomass steam gasification in fluidized bed based on machine learning method. Int. J. Hydrog. Energy. **49**, 356–370 (2024). https://doi.org/10.1016/j.ijhydene.2023.08.259
59. Hassoun, A., Cropotova, J., Trollman, H., Jagtap, S., Garcia-Garcia, G., Parra-López, C., Nirmal, N., Özogul, F., Bhat, Z., Aït-Kaddour, A., Bono, G.: Use of Industry 4.0 technologies to reduce and valorize seafood waste and by-products: a narrative review on current knowledge. Curr. Res. Food Sci. **6** (2023). https://doi.org/10.1016/j.crfs.2023.100505
60. Mohammadi, M., Rahmanifar, G., Hajiaghaei-Keshteli, M., Fusco, G., Colombaroni, C.: Industry 4.0 in waste management: an integrated IoT-based approach for facility location

and green vehicle routing. J. Ind. Inf. Integr. **36**, 100535 (2023). https://doi.org/10.1016/j.jii.2023.100535
61. Kurniawan, T.A., Meidiana, C., Dzarfan Othman, M.H., Goh, H.H., Chew, K.W.: Strengthening waste recycling industry in Malang (Indonesia): lessons from waste management in the era of Industry 4.0. J. Clean. Prod. **382**, 135296 (2023). https://doi.org/10.1016/j.jclepro.2022.135296
62. Kumar, A.: A novel framework for waste management in smart city transformation with Industry 4.0 technologies. Res. Glob., 100234 (2024). https://doi.org/10.1016/j.resglo.2024.100234
63. Zhang, A., Venkatesh, V.G., Wang, J.X., Mani, V., Wan, M., Qu, T.: Drivers of Industry 4.0-enabled smart waste management in supply chain operations: a circular economy perspective in China. Prod. Plan. Control. **34**, 870–886 (2023). https://doi.org/10.1080/09537287.2021.1980909
64. Kurniawan, T.A., Dzarfan Othman, M.H., Hwang, G.H., Gikas, P.: Unlocking digital technologies for waste recycling in Industry 4.0 era: a transformation towards a digitalization-based circular economy in Indonesia. J. Clean. Prod. **357**, 131911 (2022). https://doi.org/10.1016/j.jclepro.2022.131911
65. Zaman, A.: Waste management 4.0: an application of a machine learning model to identify and measure household waste contamination—a case study in Australia. Sustainability (Switzerland). **14** (2022). https://doi.org/10.3390/su14053061
66. Chauhan, A., Jakhar, S.K., Chauhan, C.: The interplay of circular economy with Industry 4.0 enabled smart city drivers of healthcare waste disposal. J. Clean. Prod. **279**, 123854 (2021). https://doi.org/10.1016/j.jclepro.2020.123854
67. Kanojia, A., Visvanathan, C.: Assessment of urban solid waste management systems for Industry 4.0 technology interventions and the circular economy. Waste Manag. Res. **39**, 1414–1426 (2021). https://doi.org/10.1177/0734242X21992424
68. Chicco, D., Warrens, M.J., Jurman, G.: The coefficient of determination R-squared is more informative than SMAPE, MAE, MAPE, MSE and RMSE in regression analysis evaluation. PeerJ Comput. Sci. **7**, 1–24 (2021). https://doi.org/10.7717/PEERJ-CS.623
69. Du, B., Lund, P.D., Wang, J.: Improving the accuracy of predicting the performance of solar collectors through clustering analysis with artificial neural network models. Energy Rep. **8**, 3970–3981 (2022). https://doi.org/10.1016/j.egyr.2022.03.013
70. Baashar, Y., Hamed, Y., Alkawsi, G., Fernando Capretz, L., Alhussian, H., Alwadain, A., Al-amri, R.: Evaluation of postgraduate academic performance using artificial intelligence models. Alex. Eng. J. **61**, 9867–9878 (2022). https://doi.org/10.1016/j.aej.2022.03.021
71. Hodson, T.O.: Root mean square error (RMSE) or mean absolute error (MAE): when to use them or not. Geosci. Model Dev. **7**, 1525–1534 (2022). https://doi.org/10.5194/gmd-2022-64
72. Hasanipanah, M., Jahed Armaghani, D., Khamesi, H., Bakhshandeh Amnieh, H., Ghoraba, S.: Several non-linear models in estimating air-overpressure resulting from mine blasting. Eng. Comput. **32**, 441–455 (2016). https://doi.org/10.1007/s00366-015-0425-y
73. Elmaz, F., Yücel, Ö., Mutlu, A.Y.: Predictive modeling of biomass gasification with machine learning-based regression methods. Energy. **191**, 116541 (2020). https://doi.org/10.1016/j.energy.2019.116541
74. Olatunji, O.O., Akinlabi, S., Madushele, N., Adedeji, P.A., Felix, I.: Multilayer perceptron artificial neural network for the prediction of heating value of municipal solid waste. AIMS Energy. **7**, 944–956 (2019). https://doi.org/10.3934/energy.2019.6.944
75. Yucel, O., Aydin, E.S., Sadikoglu, H.: Comparison of the different artificial neural networks in prediction of biomass gasification products. Int. J. Energy Res. **43**, 5992–6003 (2019). https://doi.org/10.1002/er.4682
76. Wen, H.T., Lu, J.H., Phuc, M.X.: Applying artificial intelligence to predict the composition of syngas using rice husks: a comparison of artificial neural networks and gradient boosting regression. Energies (Basel). **14**, 1–18 (2021). https://doi.org/10.3390/en14102932
77. Ascher, S., Wang, X., Watson, I., Sloan, W., You, S.: Interpretable machine learning to model biomass and waste gasification. Bioresour. Technol. **364**, 128062 (2022). https://doi.org/10.1016/j.biortech.2022.128062

78. Ascher, S., Sloan, W., Watson, I., You, S.: A comprehensive artificial neural network model for gasification process prediction. Appl. Energy. **320**, 119289 (2022). https://doi.org/10.1016/j.apenergy.2022.119289
79. Haq, Z.U., Ullah, H., Khan, M.N.A., Naqvi, S.R., Ahsan, M.: Hydrogen production optimization from sewage sludge supercritical gasification process using machine learning methods integrated with genetic algorithm. Chem. Eng. Res. Des. **184**, 614 (2022). https://doi.org/10.1016/j.cherd.2022.06.020
80. Pandey, D.S., Raza, H., Bhattacharyya, S.: Development of explainable AI-based predictive models for bubbling fluidised bed gasification process. Fuel. **351**, 128971 (2023). https://doi.org/10.1016/j.fuel.2023.128971
81. Yang, Y., Shahbeik, H., Shafizadeh, A., Rafiee, S., Hafezi, A., Du, X., Pan, J., Tabatabaei, M., Aghbashlo, M.: Predicting municipal solid waste gasification using machine learning: a step toward sustainable regional planning. Energy. **278**, 127881 (2023). https://doi.org/10.1016/j.energy.2023.127881

Impact of Automation on Environment and Sustainability

Anjum Qureshi

1 Introduction

The technologies are benefitting us and making our lives comfortable but at the same time, they are impacting the environment. This impact can be positive as well as negative. As the world is concerned about global warming and climate change, it has become essential to study the effect of technology to develop sustainable solutions. IoT and AI are technologies widely used to design new facilities for everyone. IoT and AI are used in many applications. Many of these applications are designed to utilize the benefits of both IoT and AI. Sustainable Development Goals (SDGs) given by the United Nations aim to achieve 17 goals by 2030. These goals ensure a peaceful and safe life for everyone. Moreover, the technologically advanced world has to achieve these SDGs by utilizing the maximum benefits of these technologies. All the SDGs are interconnected to each other. Ignoring any of them would not support achieving these SDGs within prescribed time limits. IoT and AI are being used for personal and professional applications. A few applications include health care, route prediction, disaster management, precision agriculture, smart transportation, and hospitality. All these applications affect the environment directly or indirectly. However, this effect on the environment has forced everyone to think about attaining SDGs and maintaining a circular economy. Sustainability and circular economy are interrelated. Sustainability concentrates on using resources carefully and ensuring a better future for all. At the same time, a circular economy can be achieved by minimizing waste and developing methods to extend the lifetime of the products. IoT and AI have made our lives comfortable with robust applications. But to achieve sustainability and a circular economy, it is essential to evaluate the effect of these technologies on the environment and people's health. We can hope to have a

A. Qureshi (✉)
Rajiv Gandhi College of Engineering Research and Technology, Chandrapur, India

P. Naidoo, M. Sibanda (eds.), *Measurement, Automation, and Control in Artificial Intelligence and Machine Learning*, Signals and Communication Technology, https://doi.org/10.1007/978-3-031-91802-5_6

sustainable future by balancing social, economic, and environmental factors for sustainability. The various needs of the present times should be met without compromising the requirements of future generations. We should use natural resources wisely and at the same time be conscious of their environmental impacts.

The research objectives in this chapter are as follows: (1) Explore the concepts of circular economy related to these technologies and achieving sustainability. (2) Explore the negative effects of these technologies on climate change and identify solutions for it. (3) Study how these technologies can reduce the load of environmental pollution. The findings of this chapter are based on the literature survey of the existing work on these technologies and their environmental impacts. There have been numerous studies on the environmental impact of IoT and AI. However, there is a gap in how these technologies can be used to attain a circular economy. The circular economy concepts will help reduce the negative impacts of technologies on the environment and develop solutions for making them more sustainable. This chapter has been arranged as follows: Section 2 highlights the environmental conservation aspects of IoT, the negative impacts of IoT on the environment, and how it helps achieve a circular economy and sustainability. Section 3 similarly discusses the positive and negative effects on the environment by AI and how it supports SDGs and circular economy. Section 4 discusses the research gaps and the impact of AI and IoT on forming environmental policies.

2 Internet of Things

IoT plays a crucial role in environmental applications by enabling real-time monitoring of parameters like air, water, soil, and energy consumption. IoT systems facilitate environmental conservation measures such as forest fire detection, smart agriculture, and disaster detection.

2.1 Environmental Conservation Using IoT

2.1.1 Forest Fire Detection

Forest fires are unavoidable sometimes. Consistent forest fires can cause huge damage to vegetation and wildlife. The climatic changes and rise in the global temperature may increase extended periods of fires, especially in the tropical regions. Aerial detection and ground-based detection methods are generally used to detect forest fires. These methods have some limitations:

1. The imaging system used in satellite-based detection may fail to detect the fire at the initial stages, and weather conditions like rain, clouds, and fog affect the image quality. Sometimes as these satellite-based systems operate over a larger

area, the imaging techniques may pick up industrial exhaust and volcanic eruptions as forest fires [1].
2. The Unmanned Aerial Vehicle (UAV) is considered better than satellites, but its efficiency is affected by the local meteorological conditions. Winds cause a great disturbance during landing and taking off of the UAV [2].
3. Infrared imaging identifies the heat generated after combustion in forests but cannot differentiate the thermal signatures generated by industrial exhausts and reflective surfaces. This may cause a false alarm [3].
4. Optical spectrometry [4] and Light Detection Ranging Systems (LIDAR) [5] help to overcome the problems faced by infrared imaging but these methods can work only under clean weather conditions. Their efficiency is affected by the presence of smoke and during nighttime.

IoT devices provide solutions to the challenges faced by remote detection systems. The sensor nodes are generally placed nearer to the ignition occasions and are less affected by the weather or climatic conditions. This method termed as onsite detection method is fast and more reliable. IoT-based fire detection systems offer higher reliability through real-time monitoring. Temperature, humidity, smoke, and rain sensors are basic components of an IoT-based fire detection system. Long-range communication using low-power sensors enhances the working of the IoT systems. These systems can reduce the environmental impact of forest fires by real-time monitoring and visualization of active fires [6].

2.1.2 Precision Agriculture

IoT can contribute to reducing the environmental impacts of agriculture by enabling streamlined operations. It will cause less chemical usage and reduce machine emissions. It collects real-time data on farming to know the growth of crops, weather, and climatic conditions. Real-time monitoring helps determine the exact quantity of fertilizers, pesticides, and other chemicals used in the farm. This will reduce the amount of harmful waste and thus benefit the environment [7]. Controlled use of chemicals reduces soil, land, and groundwater resource pollution. It is also helpful in preventing soil erosion. Precision or smart agriculture helps to mitigate the negative impacts of conventional farming practices. As the farmers are aware of the changes in climate and increase in population, they plan for crops and try to get maximum yields. Real-time information on the climatic conditions helps them to plan farming activities and save crops from damage. Precision farming practices minimize environmental footprints and improve crop health and productivity to ensure resilience and sustainability of agriculture [8].

2.1.3 Disaster Management

IoT can be seen as a life-saving option during disaster management. The sensor-enabled real-time detection systems help send early warning signs during fire, earthquakes, etc. IoT secures smart city infrastructure by reducing the risks of loss of humans and public property caused by disasters. IoT-based systems can manage and monitor disasters floods, volcanic eruptions, forest fires, landslides, and earthquakes. These systems can be implemented using microcontrollers, sensors, transmitters, receivers, and appropriate algorithms to collect real-time data and effectively handle ruinous situations [9]. The early warning signs and real-time environment monitoring by the IoT-based systems enable early evacuation and immediate mobilization of resources to mitigate the effects of disasters. The concerned authorities have time to plan for the safety of people. Besides this, smart infrastructure in which bridges, buildings, and similar structures are enabled with sensors, and early warning alerts help with preventive maintenance to save from infrastructure failures caused by disasters. Smart transportation by IoT is helpful to people in the selection of appropriate routes during roadblocks caused by floods, landslides, earthquakes, etc. This prevents traffic congestion and facilitates the movement of emergency response teams toward the affected areas [10].

2.2 Challenges and Negative Impacts

2.2.1 Data Privacy and Security

The voluminous data generated by the IoT devices and the several interconnected devices increase data privacy and security concerns. The environmental data is sensitive and should be protected from unauthorized access. Government policies and regulations should be strictly implemented to secure and safeguard the collected data. Prioritizing data governance frameworks and reinforcing privacy policies should be at the forefront for the government and organizations to build trust among the stakeholders. Failing to do so may hinder its responsible implementation.

2.2.2 Production and Disposal of IoT Devices

As the number of IoT devices is increasing, its production and disposal have become a challenge. Though IoT-based systems are used for environmental protection and monitoring, these devices can increase e-waste if not recycled properly. Moreover, the rapid advancements in technology have encouraged the manufacturing of new and advanced devices, and the older ones become obsolete and cannot be used as they are incompatible with the existing systems. These devices are produced using rare earth metals that increase the burden of achieving sustainability.

2.2.3 Device Limitations

The IoT devices are battery-powered. The energy consumption of these devices increases with Internet connection and voluminous data transfer. These batteries are generally low-powered, some of them are affected by climatic conditions and placed in difficult geographical locations, therefore maintenance and replacement become challenging at times.

2.2.4 Low Awareness of Sustainability

The organizations working on IoT applications related to environmental monitoring or conservation have low awareness and lack education on the negative impacts of the effective disposal of batteries and sensors. Their casual disposal into landfills can release heavy metals and toxic components into water and soil. Most manufacturers concentrate on successfully implementing their devices or applications, without any concerns of achieving sustainability.

2.3 Impact of IoT on Circular Economy and Sustainability

IoT has played a vital role in ensuring sustainability and establishing a circular economy (CE). The development of IoT devices and applications has supported connections on land and in oceans. It has helped everyone to understand the environment and wildlife.

The 17 sustainable development goals (SDGs) of the United Nations are interlinked with each other and IoT [11]. Some of the goals that are strongly correlated with IoT are:

1. SDG 6 (Clean Water and Sanitation) can be achieved by enabling monitoring of water quality, using smart water meters to control the usage of water in households;
2. SDG 7 (Affordable and Clean Energy) can be attained through smart and energy-efficient buildings that use IoT-based devices. Some examples are smart energy meters to control and inform the users of their daily energy usage, smart lights that can control their intensity depending on natural illumination in the room, and automatic temperature adjustments based on weather conditions.
3. SDG 9 (Industry Innovation and Infrastructure) can be achieved through real-time monitoring of machines and devices in the industry by using IoT. This facilitates preventive and predictive maintenance and thus avoids wasting time, money, and human labor when the industry might come to a standstill due to machine failures. Moreover, the interconnected devices and real-time data collection enable manufacturing optimization to lower energy consumption and improve productivity.

4. SDG 11 (Sustainable Cities and Communities) can be achieved through IoT-enabled applications of smart cities. Some of these applications are smart waste management that uses smart bins to alert the municipal authorities when the bin is full, smart traffic lights that can control on and off timing according to the traffic, smart parking systems, smart transportation that enables route navigation to avoid traffic congestion, and finding facilities like hospitals, petrol pumps, etc. Smart transportation saves fuel and is an important step toward a cleaner environment. IoT helps efficient energy management by using smart street lighting systems that automatically switch on lights for the arrival of persons and vehicles.
5. SDG 13 (Climate Action) is achieved through IoT applications like environmental monitoring, tracking emissions from the industries, monitoring soil health, and optimizing irrigation for precision agriculture, disaster management preparedness, and smart and energy-efficient buildings.

The correlation of IoT with a few SDGs mentioned above indicates that IoT is an important element of the circular system. It enables smart energy usage and reduces emissions and pollution. Real-time monitoring enables predictive and preventive maintenance to facilitate easy replacement of parts, thereby promoting reutilization and repair and avoiding unnecessary new equipment. It supports optimizing manufacturing operations and fast and efficient outflow of goods using robotic pickers and automated guided vehicles, which are a few applications that support circular production technologies. These features provide better visibility of the supply chains. The organizations can plan for production and manufacturing to fulfill the demands of circular supply chains [12]. The real-time detection for the resources inventory is implemented using IoT by the Industry 4.0 generation [13]. Minimization of waste during the lifecycle of a product fosters avenues to optimize its reuse and recycling. IoT enables meticulous planning for efficient resource utilization, supports renewable resources, and facilitates the development of circular business models [14]. IoT enables smart manufacturing and sustainable supply chain management by real-time monitoring and tracking capabilities. It thus expedites the formulation of measures to decrease energy consumption and avoid exploitation of resources. This ensures lower carbon footprints in various industries [15]. IoT promotes green and sustainable development by supporting the designing of scheduling and logistics delivery models for intelligent production [16]. IoT uses sensors to collect information about post-consumption products, thereby enabling recycling and repairing [17]. The organizations use sensors in their products to monitor the maintenance requirements and thus provide high-quality service to their customers. This helps organizations to improve the performance and life span of the products and supports reduction, reuse, and recycling [18]. Some organizations use IoT to collect the satisfaction and reactions of consumers to their past products. The designers use this data to improve the product quality [19]. IoT promotes circular business models by allowing reduction, reuse, recycling, repair, redesign, and remanufacture called the 6Rs. These actions increase the usage of the products and their lifetime [20].

3 Artificial Intelligence

AI can play a pivotal role in reducing GHG emissions and enhance the efficiency of applications in various fields like healthcare, education, communication, transportation, etc. This section will discuss how AI will contribute to sustainability and the hidden environmental impacts of AI.

3.1 Improving Sustainability

An analysis of AI indicates that it can facilitate 134 targets of the UN SDGs while it may obstruct the achievement of 59 targets [21]. AI will contribute to attaining sustainability by reducing energy consumption, having smarter transportation in cities, enhancing the performance of precision agriculture, and reducing emissions.

3.1.1 Reduced Energy Consumption

One of the key challenges to be faced with increased urbanization is fulfilling the electricity and power demand for its residents. It is not feasible to increase the power generation capabilities beyond a limit but the requirements can be fulfilled by adopting energy-efficient techniques. One of these techniques is smart grids that use digital technologies to meet the supply and demand of electricity and minimize costs without affecting the reliability and stability of the grid [22]. AI-based smart grids can help to reduce electricity usage by optimizing power generation and local demand. Incorporating renewable energy sources in smart grids helps to overcome voltage fluctuations and energy losses as observed in traditional grids. The availability of renewable energy sources like solar and wind may vary according to weather. ML technique can be used to forecast the availability of resources to ensure uninterrupted integration into the smart grid [23]. These models analyze the historical data and use the real-time data to deal with the uncertainties in production and complexities of the smart grids. Deep neural networks are widely used to forecast wind energy [24], and hybrid methods that combine deep neural networks, recurrent networks, and long short-term memory networks are used for predicting solar energy [25]. ML algorithms also help to analyze the data collected by the sensors in smart buildings to optimize their energy usage for activities like heating, ventilation, lighting, air conditioning, etc. It makes the buildings energy efficient and decreases carbon emissions [26].

3.1.2 Improved Transportation

Traffic congestion and environmental pollution are becoming a critical issue with the rapid increase in urban population. AI can make traditional transportation systems more sustainable by predicting traffic patterns and optimizing routes. Real-time analysis using these algorithms can regulate the public transportation system by using dynamic scheduling to ensure well-timed arrival and departure of the vehicles. This promotes eco-friendly modes of transport where people prefer using shared transport services and reap the benefits of an interconnected sustainable environment.

Autonomous Vehicles (AV) use AI to improve safety and decrease the chances of road accidents [27]. ML algorithms enhance the working of AV by using reinforcement learning in AV controllers to manage traffic interactions and minimize the energy consumption of AVs [28].

3.1.3 Sustainable Precision Agriculture

ML algorithms enhance the performance of precision agriculture by providing farmers with valuable information on matters related to agriculture like crop health, irrigation, and soil. The real-time data collected by the sensors, satellites, and drones is used to estimate the requirement of fertilizers, pesticides, and water for the crops. Predictive analytics using AI has additional benefits like forecasting crop yield and detecting crop diseases to enable farmers to take necessary steps to minimize losses [29]. Some of the ML models used in precision agriculture are:

Convolutional Neural Networks (CNNs) are used to predict crop yields using data acquired from the UAVs [30], random forest (RF), and support vector machine (SVM) for detecting groundwater levels [31]. Quantum ML is an effective technique for detecting crop diseases and adopting mechanized farming activities. AI-based equipment like autonomous tractors and robotic harvesters reduces the ecological footprint of smart agriculture [32].

3.1.4 Climate Change

ML techniques are extensively used to predict the demand and supply of energy and the availability of renewable energy sources like sun and water. Supervised ML techniques and fuzzy logic are used to analyze historical data and the data obtained as videos and photos [33]. ML applications help to reduce fossil fuel emissions and mitigate methane leaks in natural gas pipelines. Convolutional Neural Networks (CNNs) automatically detect methane leaks in natural gas pipelines by training the CNN with images of methane leaks [34]. These ML algorithms can be an additional benefit to the environment by reducing emissions. Using ML algorithms will cause less reliance on carbon-intensive materials. It will help develop industrial processes that use lower carbon energy and structures with reduced usage of cement and steel [35].

3.2 *Negative Impacts of AI on the Environment*

As discussed in the previous section, AI can contribute to protect the environment and attain sustainability. However, some hidden aspects should be addressed to minimize the environmental impact caused by AI.

3.2.1 Carbon Footprint

AI and ML algorithms are being used increasingly in diverse applications that are helping to reduce emissions and contribute to sustainability. However, as more people have started using AI due to its benefits, researchers are raising concerns over the power consumption of AI and its carbon footprint.

A report by Harvard Business Review states that training an AI model can consume thousands of megawatts of power. It will furthermore release tons of carbon into the environment [36]. Open AI researchers have specified that the computing power requirement of training AI models doubles every 3.4 months. With this rate, it is expected that by 2040 nearly 14% of the global emissions will originate from the data centers and communication networks. The University of Massachusetts during recent research shared that 626,000 pounds of carbon dioxide are generated while training AI models. This amount of carbon dioxide is five times the emissions caused by a car in a lifetime and equal to three hundred round-trip flights from New York to San Francisco [37]. With the increasing use of AI, new models are being developed and need regular upgrades. The carbon footprints of the latest models are always higher than the older versions. In such circumstances, regular upgrades will incur large energy costs.

3.2.2 Increase of E-Waste

E-waste produced by AI technology has become an environmental challenge. This e-waste contains hazardous chemicals like mercury, lead, and cadmium that contaminate soil, water, and land. The harm caused by these chemicals can be minimized by developing novel techniques for recycling and managing e-waste. The negative effect of e-waste on the environment can be achieved by following ethical practices for waste disposal and having strict laws for those who violate these practices. The construction of data centers for AI infrastructure will require more land. It will disrupt wildlife and may seem to be a threat to biodiversity and environmental resilience.

3.2.3 Increased Water Consumption

The data centers need large volumes of water to cool down the heat generated by the IT equipment in the data center. Conventional air conditioning systems are used to cool the data centers but their energy consumption is very high. Moreover, heat rejection by the small fans in the cooling system increases temperature, thereby making cooling difficult for many people [38]. However, the air blown by the large fans causes evaporation. An organization should select an appropriate and low-power-consuming cooling method for the data centers.

3.3 Transformation to Circular Economy

AI has positive and negative impacts on the environment. However, its current capabilities and future potential can play a pivotal role in developing a circular economy. Creating AI applications and their implementation can be troublesome at times. We need experts to make the training data and develop an algorithm to yield results that would be used easily and beneficial for many. The success of AI algorithms however depends on the quality of data. Poor quality of data may result in unsatisfactory outputs. AI can speed up the formation of a circular economy. It can facilitate the designing and development of new circular products and help the conventional ones to become circular.

AI technologies can help industries unleash the following promising opportunities for heading toward a circular economy: (1) produce better and new circular commodities, (2) facilitate functioning of circular business model, and (3) enhance the existing frameworks to assure the availability of circular products and models [39].

3.3.1 Better and New Circular Commodities

The circular economy calls for innovation that helps design products with higher utilities and a longer life. Moreover, these products and commodities should enable reusing, recycling, and repairing to allow longer usage and reduce waste. These products should be designed considering factors like disassembling, upgrading, and recycling. AI technology can be a great help to designers in selecting appropriate components for their new design. They can leverage the feedback process and AI-generated suggestions to test and improve their designs. This would help in making better products in a shorter time. The circular economy also focuses on developing designs that would be environmentally friendly. These products should be free from materials that harm the environment. AI can help in predicting the toxicity of materials. It can also be used to develop a systematic way of producing and testing new metal alloys.

3.3.2 Facilitate Circular Business Model

Marketing, sales, customer support, price, and logistics are some business functions that help to establish circular economy principles. Successful, profitable, and innovative circular business models can be developed by planning how a returned product can be used in the next cycle. It requires the design of effective techniques for reusing and recycling. This can be attained by blending aspects like the present condition of the product and its market scenario and by planning to make the circular products acceptable among the people amid existing linear products. Designing a circular model by considering all these factors can be done by analyzing a large quantity of data for products and consumers. AI can be a great analytical tool and help in decision-making. AI algorithms can help manufacturers to set competitive prices for their products. Moreover, it can help to study the fluctuation in demand and supply of used products and their return. AI can create circular value for the currently existing business models. It can cut back inventory levels without affecting the consumer's demand. All these would help in reducing waste. As the products are acceptable and used by people, there will be fewer unsold products and their costs will be reduced.

3.3.3 Enhancing Existing Frameworks

One of the key features of a circular economy is to have products that can be used for longer periods. Avoiding disposal of waste to maximize its usage time can be achieved by developing a comprehensive structure to collect, sort, treat, separate, and redistribute. AI can play a pivotal role in making valuable products from the used, to ensure a circular flow of products. For example, visual recognition techniques based on AI use cameras and sensors to use intelligent robots to sort and separate waste. Presort and separation help to achieve better levels of recovery. Improved identification of materials for reuse and remanufacture will enhance the quality of recovered materials. AI algorithms can analyze the real-time data of sensors to analyze waste and improve the efficiency of waste sorting methods.

3.4 Effects on SDGs

AI-powered technologies can help achieve SDGs related to the environment and its conservation. These goals are SDG 13 (Climate action), SDG 14 (life below water), and SDG 15 (life on land). AI can act as a facilitator in meeting 25 targets for these SDGs [40]. Real-time detection on large-scale interconnected networks has enabled AI to achieve these targets [41]. AI supports SDG 13 by understanding the changes in climate and predicting its impact. It also supports the integration of renewable energy sources to have more energy-efficient systems [42]. AI algorithms can automatically identify the likely oil spills that can help reduce marine pollution and save life underwater [43]. ML algorithms have proved to be successful in conserving coastal and marine areas. These algorithms utilize various parameters to provide favorable solutions that help to identify areas for implementing conservation measures [44]. Neural networks and object-oriented techniques in AI are used to classify vegetation into different types using satellite images. These algorithms can produce fast results by processing several images in very little time. The classification of the vegetation cover using AI can help deal with issues like deforestation and desertification and restore the degraded land and soil. Analyzing the images of desertification would help to identify the pattern of deforestation and its major drivers. However, it would use the collected information to plan and make decisions for avoiding deforestation and desertification [45]. Similarly automated processes can be used to collect data on biodiversity. These algorithms will support studying the behavioral pattern of any species and generate notifications or alerts for unusual incidents.

4 Research Gaps and Impact on Environment Policies

Though AI is increasingly considered an enabler to achieving SDGs, organizations and researchers should try to concentrate on preventing malfunctioning and hacking of these systems. Research on ensuring safety for AI should be encouraged to make these systems more beneficial for all. Moreover, there is a need to raise awareness of the risks associated with the failure of AI systems. Society has become dependent on these AI-based devices and applications due to speed, accuracy, and overall comfort compared to the conventional methods. Moreover, the studies on SDGs and their targets using AI are done using limited datasets and prototypes. Evaluating this information in real time and measuring its impact on a broader scale become challenging. As the demand for AI is increasing and the requirements of people are changing, there is a need to develop applications that would be novel and could be tailored to satisfy the consumers. At the same time, morality, adaptability, accountability, and sustainability should be the primary concerns of those developing new AI applications for the people. The research community is set to give immediate and positive results using AI in the urge to achieve the SDG goals. However, very few are concerned about the long-term detrimental effects of AI-based systems. The lack of evaluation methodologies and the haste to publish only positive results prevent people from participating in research on the devastating results of AI on the planet. As more people are interested in researching AI, these projects will receive more funding opportunities than other technologies. This would create an inequality where people would prefer working on AI to satisfy their commercial interests. Moreover, people would prioritize AI-based technologies over the SDGs to maximize economic and commercial viability. Organizations must encourage people to maintain honesty and integrity to avoid harming the environment. AI applications are being used and designed by technologically advanced nations. Many developing and low-income countries are far from the knowledge and usage of these applications. A digital divide thus still exists. A part of the world may solve environmental issues using AI but others may still struggle to use AI. In such circumstances, achieving environmental-related SDGs for everyone seems challenging. Researchers and funders should focus on developing applications to be used by low-income countries. Moreover, these applications should be tailored to the local environment. This would create a better understanding and increase the adaptability of AI applications to achieve SDGs and reduce the burden of climate change and global warming.

Many researchers and designers have neglected the impact of using IoT on the environment. They have been trying to communicate the advantages of IoT applications. The consumers are unaware of the environmental impact of IoT as they have not received appropriate information. IoT makes energy-efficient buildings by monitoring their energy usage. However, there have been minimal discussions on the energy consumption of IoT devices. For example, though we say that the energy consumption of long-range (LoRa) wireless technologies and RFID tags is less, as the number of such devices increases in IoT applications, their total energy consumption can be of high value. The researchers concentrate on reducing energy consumption using IoT applications but the energy consumption of the devices used in implementing an application is neglected. Moreover, the carbon emissions from the sensors and batteries used in the IoT are hardly discussed. There is less information available on the energy calculation of the IoT nodes. Besides that, data on how IoT promotes a circular economy and circular business models is limited. Moreover, a lack of standardization in implementing IoT across business models and limited regulations for data ownership issues obstruct the implementation of the circular economy. IoT applications like those used for smart transportation, smart buildings, route navigation, and smart healthcare collect personal data. Consumers are exposing their data in exchange for products and services but are unaware if their data is secure. They have no idea if their data is shared with third parties without their consent.

IoT and AI can emerge as robust solutions to detect environmental violations and support proactive methods to protect the environment. IoT uses sensor networks for real-time monitoring of parameters like air quality and harmful emissions to predict the effect of various activities on the environment. It can detect the presence or leakage of hazardous gases and generate alerts for the stakeholders to prevent any harm to the people. Additionally, the data analytical capabilities of AI will be instrumental in improving decision-making for better environment management. However, the widespread adoption of IoT and AI as enablers of environment protection will be strengthened by overcoming the data management and connectivity challenges. Several industries rely on fossil fuels for their production and other related activities. It is not feasible for the persons appointed by the environmental regulatory organizations to regularly evaluate and inspect the violators. IoT can be used for collecting data and AI to analyze the gathered data. This would help to predict which organizations or industries are likely to be potential violators. IoT and AI can enable finding these violators without visiting those places. However, it will help design more efficient policies with limited resources. Besides that, AI can be used to detect global climate changes through satellite monitoring. AI algorithms should be handled responsibly. It can create fair and accurate systems when trained using quality data. Human biases should be avoided, and the AI algorithms should be trained on appropriate data to ensure environmental justice. All people should be protected from the negative consequences of the environment caused by industrial and commercial operations. The capabilities of IoT and AI can be used to frame policies and practices that can prevent the sources of toxic wastes and emissions from being located away from the people. This will ensure better health for everyone. However, sustainability can be attained by making IoT and AI environment-friendly. These technologies should leverage green aspects to support environmental conservation. Green IoT and Green AI are emerging concepts that concentrate on developing new software and hardware without leaving any negative impact on the environment. These new products or services should be reusable and have minimum computation time and memory requirements.

5 Conclusion

This chapter has tried to highlight the importance of IoT and AI in protecting the environment. It also tries to bring out their negative effects on the environment. Though these technologies hurt the environment, their capabilities of achieving SDGs and circular economy make them one of the most acceptable technologies. However, looking at the negative aspects, it can be said that more research is required to have greener IoT and AI. The environmental benefits and drawbacks should be analyzed carefully to protect the environment and ensure a circular economy. The assessments of advantages and disadvantages would provide fundamental knowledge for the researchers interested in circular development. The research gaps in this regard have also been identified. These research gaps should be addressed to achieve sustainability and environmental protection using IoT and AI.

References

1. Milz, M.: Study on Forest Fire Detection with Satellite Data, p. 23. Luleå Tekniska Universitet (2013)
2. Merino, L., Caballero, F., Martínez-de-Dios, J.R., Maza, I., Ollero, A.: An unmanned aircraft system for automatic forest fire monitoring and measurement. J. Intell. Robot. Syst. **65**(1–4), 533–548 (2012). https://doi.org/10.1007/s10846-011-9560-x27
3. Arrue, B.C., Ollero, A., Matinez de Dios, J.R.: An intelligent system for false alarm reduction in infrared forest-fire detection. IEEE Intell. Syst. **15**(3), 64–73 (2000). https://doi.org/10.1109/5254.846287
4. Dennison, P.E.: Fire detection in imaging spectrometer data using atmospheric carbon dioxide absorption. Int. J. Remote Sens. **27**(14), 3049–3055 (2006). https://doi.org/10.1080/01431160600660871

5. Lavrov, A., Utkin, A.B., Vilar, R., Fernandes, A.: Evaluation of smoke dispersion from forest fire plumes using lidar experiments and modelling. Int. J. Therm. Sci. **45**(9), 848–859 (2006). https://doi.org/10.1016/j.ijthermalsci.2006.01.003
6. Sharma, A., Nayyar, A., Singh, K.J., et al.: An IoT-based forest fire detection system: design and testing. Multimed. Tools Appl. **83**, 38685–38710 (2024). https://doi.org/10.1007/s11042-023-17027-9
7. Dhanaraju, M., Chenniappan, P., Ramalingam, K., Pazhanivelan, S., Kaliaperumal, R.: Smart farming: Internet of Things (IoT)-based sustainable agriculture. Agriculture. **12**, 1745 (2022). https://doi.org/10.3390/agriculture12101745
8. Meena, A.K., Meena, R.N., Meena, A.K.: Use of precision agriculture for sustainability and environmental protection. In: Book Advances in Agronomy. AkiNik Publications (2019)
9. Sharma, K., Anand, D., Sabharwal, M., Tiwari, P.K., Cheikhrouhou, O., Frikha, T.: A disaster management framework using Internet of Things based interconnected devices. In: Mathematical Problems in Engineering. Wiley Online Library (2021)
10. Ding, Z., Jiang, S., Xinrun, X., Han, Y.: An Internet of Things based scalable framework for disaster data management. J. Saf. Sci. Resil. **3**(2), 136–152. ISSN: 2666-4496 (2022). https://doi.org/10.1016/j.jnlssr.2021.10.005
11. Phillipe, J., Charman, R.: IoT Enabling Sustainability for Circular Economy. Software AG (2022). https://blog.softwareag.com/circular-economy-sustainability/
12. Voulgaridis, K., Lagkas, T., Angelopoulos, C.M., Nikoletseas, S.E.: IoT and digital circular economy: principles, applications, and challenges. Comput. Networks. **219**, 109456. ISSN: 1389-1286 (2022). https://doi.org/10.1016/j.comnet.2022.109456
13. Mohammadian, H.D.: IoE—a solution for energy management challenges. In: 2019 IEEE Global Engineering Education Conference (EDUCON) (2019)
14. Nižetić, S., Šolić, P., López-de-Ipiña González-de-Artaza, D., Patrono, L.: Internet of Things (IoT): opportunities, issues and challenges towards a smart and sustainable future. J. Clean. Prod. **274**, 122877 (2020). https://doi.org/10.1016/j.jclepro.2020.122877
15. Ghoreishi, M., Happonen, A.: The case of fabric and textile industry: the emerging role of digitalization, internet-of-things and Industry 4.0 for circularity. In: Proceedings of Sixth International Congress on Information and Communication Technology, vol. 216, pp. 189–200 (2022). https://doi.org/10.1007/978-981-16-1781-2_18
16. Liao, W., Wang, T.: A novel collaborative optimization model for job shop production–delivery considering time window and carbon emission. Sustainability. **11**(10), 2781 (2019). https://doi.org/10.3390/su11102781
17. Gligoric, N., Krco, S., Hakola, L., Vehmas, K., De, S., Moessner, K., Jansson, K., Polenz, I., Van Kranenburg, R.: Smart tags: IoT product passport for circular economy based on printed sensors and unique item-level identifiers. Sensors. **586**(3) (2019). https://doi.org/10.3390/s19030586
18. Nobre, G.C., Tavares, E.: Scientific literature analysis on big data and Internet of Things applications on circular economy: a bibliometric study. Scientometrics. **111**, 463–492 (2017). https://doi.org/10.1007/s11192-017-2281-6
19. Li, C.Z., Chen, Z., Xue, F., et al.: A blockchain- and IoT-based smart product-service system for the sustainability of prefabricated housing construction. J. Clean. Prod. **286**, 125391 (2021). https://doi.org/10.1016/j.jclepro.2020.125391
20. Ding, S., Tukker, A., Ward, H.: Opportunities and risks of Internet of Things (IoT) technologies for circular business models: a literature review. J. Environ. Manag. **336**, 117662. ISSN: 0301-4797 (2023). https://doi.org/10.1016/j.jenvman.2023.117662
21. Vinuesa, R., Azizpour, H., Leite, I., Balaam, M., Dignum, V., Domisch, S., Felländer, A., Langhans, S.D., Tegmark, M., Fuso Nerini, F.: The role of artificial intelligence in achieving the sustainable development goals. Nat. Commun. **11**(1), 1–10 (2020)
22. Wietfeld, C., Cardenas, A.A., Chen, H.-H., Popovski, P., Wong, V.W.S.: Smart grids. IEEE Wirel. Commun. **24**, 8 (2017). https://doi.org/10.1109/MWC.2017.7909091

23. Aslam, S., Herodotou, H., Mohsin, S.M., Javaid, N., Ashraf, N., Aslam, S.: A survey on deep learning methods for power load and renewable energy forecasting in smart microgrids. Renew. Sustain. Energy Rev. **144**, 110992 (2021)
24. Lin, K.-P., Pai, P.-F., Ting, Y.-J.: Deep belief networks with genetic algorithms in forecasting wind speed. IEEE Access. **7**(2019), 99244–99253 (2019)
25. Abdel-Nasser, M., Mahmoud, K.: Accurate photovoltaic power forecasting models using deep LSTM-RNN. Neural Comput. Appl. **31**, 2727 (2019)
26. Zhang, M., Kafy, A.-A., Xiao, P., Han, S., Zou, S., Saha, M., Zhang, C., Tan, S.: Impact of urban expansion on land surface temperature and carbon emissions using machine learning algorithms in Wuhan, China. Urban Clim. **47**, 101347 (2023)
27. Bojarski, M., Del Testa, D., Dworakowski, D., Firner, B., Flepp, B., Goyal, P., Jackel, L.D., Monfort, M., Muller, U., Zhang, J., et al.: End to end learning for self-driving cars (2016). arXiv preprint arXiv:1604.07316
28. Wu, C., Kreidieh, A., Vinitsky, E., Bayen, A.M.: Emergent behaviors in mixed-autonomy traffic. In: Conference on Robot Learning, pp. 398–407. PMLR (2017)
29. Sharma, R., Kamble, S.S., Gunasekaran, A., Kumar, V., Kumar, A.: A systematic literature review on machine learning applications for sustainable agriculture supply chain performance. Comput. Oper. Res. **119**, 104926 (2020)
30. Nevavuori, P., Narra, N., Lipping, T.: Crop yield prediction with deep convolutional neural networks. Comput. Electron. Agric. **163**, 104859 (2019)
31. Pham, Q.B., Kumar, M., DiNunno, F., Elbeltagi, A., Granata, F., Islam, A.R.M.T., Talukdar, S., Nguyen, X.C., Ahmed, A.N., Anh, D.T.: Groundwater level prediction using machine learning algorithms in a drought-prone area. Neural Comput. Appl. **34**(13), 10751–10773 (2022)
32. Genemo, M.: Quantum convolutional neural network for agricultural mechanization and plant disease detection. In: International Conference on Image Processing and Capsule Networks, pp. 225–237. Springer (2023)
33. Hong, T., Fan, S.: Probabilistic electric load forecasting: a tutorial review. Int. J. Forecast. **32**(3), 914–938 (2016)
34. Wang, J., Tchapmi, L.P., Ravikumar, A.P., McGuire, M., Bell, C.S., Zimmerle, D., Savarese, S., Brandt, A.R.: Machine vision for natural gas methane emissions detection using an infrared camera. Appl. Energy. **257**, 113998 (2020)
35. Kazi, R.H., Grossman, T., Cheong, H., Hashemi, A., Fitzmaurice, G.W.: DreamSketch: early-stage 3D design explorations with sketching and generative design. UIST. **14**, 401–414 (2017)
36. Ren, S., Wierman, A.: The uneven distribution of AI'S environmental impacts. Harvard Business Review (2024). https://hbr.org/2024/07/the-uneven-distribution-of-aisenvironmentalimpacts#:~:text=Even%20putting%20aside%20the%20environmental,hundreds%20of%20tons%20of%20carbon
37. Kanungo, A.: The Green Dilemma: Can AI Fulfil Its Potential Without Harming the Environment? (2023). https://earth.org/the-green-dilemma-can-ai-fulfil-its-potential-without-harming-the-environment/
38. Ziegler, M.T.: The World's AI Generators: Rethinking Water Usage in Data Centers to Build a More Sustainable Future
39. Ellen MacArthur Foundation: Artificial Intelligence and the Circular Economy: AI as a Tool to Accelerate the Transition (2019). https://www.ellenmacarthurfoundation.org/artificial-intelligence-and-the-circular-economy
40. Vinuesa, R., Azizpour, H., Leite, I., et al.: The role of artificial intelligence in achieving the sustainable development goals. Nat. Commun. **11**, 233 (2020). https://doi.org/10.1038/s41467-019-14108-y
41. World Economic Forum (WEF): Fourth Industrial Revolution for the Earth Series Harnessing Artificial Intelligence for the Earth. World Economic Forum (2018)
42. Mathews, R.: AI'S Impacts of the SDGs. Change Oracle (2024). https://changeoracle.com/2024/02/26/ai-and-the-sdgs/

43. Keramitsoglou, I., Cartalis, C., Kiranoudis, C.T.: Automatic identification of oil spills on satellite images. Environ. Model. Softw. **21**, 640–652 (2006)
44. Beyer, H.L., Dujardin, Y., Watts, M.E., Possingham, H.P.: Solving conservation planning problems with integer linear programming. Ecol. Model. **328**, 14–22 (2016)
45. Mohamadi, A., Heidarizadi, Z., Nourollahi, H.: Assessing the desertification trend using neural network classification and object-oriented techniques. J. Fac. Istanb. Univ. **66**, 683–690 (2016)

Artificial Intelligence (AI) and Machine Learning (ML) in Control Valve Systems

Puramanathan Naidoo

1 Introduction

The FCE in a process control loop is either a control valve or VSD varying the speed of a motor driving a pump. The FCE regulates the amount of fluid in the process loop to maintain the set point of the controller. The more common FCE is the control valve. Traditionally, the FCE is operated by receiving a signal from the controller. This signal is either automatically generated through the control algorithm within the controller or manually through pre-set values by the operator. However, technological advancements make it possible to introduce AI and ML to valve management systems that are data driven. Since control valves are purely mechanical devices, as one of the instruments in an electronic loop, AI and ML bring unique precision capabilities to control valve position, enhancing reliability.

AI and ML facilitate a computer system to receive data, store the data, learn from the data, and make decisions based on human reasoning, in the absence of a specific program, for various tasks. AI is a concept where plant data is analysed in real time and stored, and ML algorithm can analyse the historical sensor data to predict any anomalies in the valve that require online or offline attention. Control valves can now adjust to real-time process conditions, detect maintenance requirements to prevent breakdowns, and enhance operations, in the absence of manual intervention. This reduces downtime and allows for adaptability of this mechanical device into smarter, unified systems. AI and ML technologies transform process control valves from simple mechanical components into intelligent devices, in line with 4IR requirements.

P. Naidoo (✉)
Mangosuthu University of Technology, Durban, South Africa
e-mail: pnaidoo@mut.ac.za

P. Naidoo, M. Sibanda (eds.), *Measurement, Automation, and Control in Artificial Intelligence and Machine Learning*, Signals and Communication Technology, https://doi.org/10.1007/978-3-031-91802-5_7

2 Overview of Control Valve Systems

Since the control valve is a purely mechanical device, provision needs to be made for all internal moving components to be adapted to advanced industrial process automation and control systems. Valves originally started as early manual systems like a normal tap at home. It required an operator to open or close, to increase or decrease the amount of process medium flowing through it, respectively. The reason why the control valve is an extremely important instrument in a process loop is that it serves the purpose to primarily regulate the process medium in all process loops. Irrespective of whether the process variable being measured is level, pressure, flow, pH, etc., it is the control valve that regulates the process medium, for the desired process variable to be achieved. Control valves have evolved to be available in various forms physically for their respective applications and industries and sophisticated in that digitally controlled valves can now be adapted to 4IR, AI and ML systems. Control valve systems have evolved alongside industrial needs and technological progress. The following subsections provide an overview of control valves to provide the reader with insight of why and how they can function in 4IR, AI and ML systems [1, 2].

2.1 Why a Control Valve Instead of a VSD as the FCE

A valve positioner, like a VSD, is an essential instrument in the process control loop that cannot function on its own. A valve positioner functions only with a control valve, just as a VSD that functions only with a motor that turns a pump, an agitator, amongst others. Although both instruments regulate the flow of the process medium to maintain the required process variable being measured, process control valves are most often used over VSDs for many important reasons as discussed below:

2.1.1 Precision in Flow Control

The primary reason for choosing process control valves over VSDs is their high level of accuracy in adjusting the amount of fluid or gas for all variables being measured. VSDs are effective but they regulate motor speed to regulate the process medium like pumps whilst process control valves regulate flow rates directly in the process line, thereby facilitating exact adjustments of level, flow rate, pressure, or temperature.

2.1.2 Quick Response Time and Stability

Process control valves respond faster, especially if a valve positioner is used to stabilise process variable as expected. VSDs regulate the process medium by varying motor speed, introducing a delay due to the dynamics of the motor, like ramping up or down, and the connected pump, which has its own dynamics. This arrangement creates a restriction for rapid flow process adjustments. Control valves receive direct transmission signals instantly, increasing response time in dynamic processes, thereby maintaining stability of the process. This level of stability is critical on boiler steam control or pressure control in the petroleum industry, where even minimal process variations impact on quality or, more importantly, safety.

2.1.3 Suitability for High-Pressure and High-Temperature Environments

Process control valves have another distinct advantage over VSDs due to them being able to withstand high-pressure or high-temperature conditions. VSDs are electrical devices and as such cannot withstand severe operating conditions like a control valve. Although VSDs can be located in enclosures, the limitations do persist, due to their electrical nature, resulting in limitations of performance and longevity. Process control valves are manufactured in accordance with the process application and environment, produced from a wide variety of materials, e.g. carbon steel, stainless steel, etc.

2.1.4 Cost-Effectiveness for Certain Applications

Like a process control valve, a VSD is not the primary driver of fluid flow. Each motor driving a pump or fan requires a VSD each, to regulate the process medium, which becomes an expensive exercise, especially in complex processes. Control valves offer a concentrated and efficient practice to regulate processes, providing a variety of options to suit the application. The installation and maintenance costs of VSDs can be high, whilst control valves can reduce the cost of maintenance, with lower operational cost.

2.1.5 Compatibility with Diverse Fluid Types and Flow Characteristics

Process control valves can be customised for various processes, e.g. fluids, liquids, gases, and slurries, to regulate different viscosities and densities. VSDs are normally restricted to cleaner fluids or gases due to the physical dynamics of pumps and electrical dynamics of motors. VSDs are not appropriate for applications such as wastewater, slurry, abrasive, or corrosive processes as they are more susceptible to blocking and damage. Various valve designs like diaphragm, ball, plug and seat valves are manufactured for the various types of processes, expanding to designs for different flow characteristics.

2.1.6 Enhanced Safety and Reliability in Critical Processes

Where reliability and safety are key features in process applications, process control valves serve as the ideal option due to their rigid design and reliability to operate without compromising process requirements. Being mechanical devices, they are not as vulnerable to electrical failures may influence VSDs. Furthermore, in emergency applications where a process medium needs to shut off and remain physically shut control valves offer the fail-safe condition to shut off, improving system and human safety.

2.2 Early Development of Control Valves in the Nineteenth Century

The first mechanical control systems commenced in the eighteenth and nineteenth century industrial revolution, seeing the growth of steam engines, making it necessary to regulate variables like steam pressure and flow rate, for safety reasons. This phase saw the introduction of the simple manual valve. This made it possible for operators to regulate the steam flow rate manually, to monitor the steam pressure on a pressure gauge, which was also invented in the 1800s. These earlier valves were restricted to open/close which were developed to offer the regulating option. However, these valves were further developed giving rise to more advanced regulatory mechanisms but remained totally dependent on the human element. The need for the human element, referred to as open loop control, resulted in limiting factors like effective control and safety, with the need for operators to monitor the process and make manual adjustments depending on process conditions [3].

2.3 The Control Valve

With the advent of pneumatic systems in the twentieth century brought advancements in pneumatic (air-powered) valves making it possible for valves to be controlled automatically, referred to as closed-loop control, in which human intervention was not required to regulate the process variable. An air compressor delivers the required air pressure to the pneumatic valve, to operate. There are two types of control valves, viz. Air-to-Open (ATO) and Air-to-Close (ATC). Due to the valve being a mechanical device, it is the only instrument that requires no supply. The two forces that determine the position of the control valve are:

- Spring tension keeps it in the normal position, e.g. an ATC valve is opened in its normal position
- Variable air signal (20–100 kPa) takes it into the opposite position proportionately, e.g. an ATC valve opens as the air signal increases by overcoming the spring tension [4]

The important feature of the pneumatic control valve, specifically being mechanical, is due to the spring tension keeping it locked in either the open or close position, until an increasing air signal is applied. This gives rise to a control valve being one of the following types when installed in a process application:

- Fail-Safe Open (FSO), where an ATC valve will be forced to return to its fully opened position under any unusual process conditions, for safety reasons or specific process requirements
- Fail-Safe Close (FSC), where an ATO valve will be forced to return to its fully closed position under any unusual process conditions

The cross section of an ATO pneumatic valve can be seen as a purely mechanical device, as represented in Fig. 1.

A summary of the operation appears below, with reference to Fig. 1:

- The valve is shut in its normal position
  - It is the spring tension that keeps it shut
- u is normally connected at *compressed air in*
  - This valve is fitted with a valve positioner (discussed in Sect. 2.3)
- There is a constant air supply of 150 kPa at p_s
- An input pneumatic signal u of 20–100 kPa is applied to the valve positioner
- An output pneumatic signal $p1$ of 20–100 kPa is transmitted from the valve positioner to the valve at *compressed air in*
- $p1$, when applied to the valve, causes an increase in pressure at $p2$
- As the pressure increases at $p2$, it results in the spring being compressed at F_a, which results in the stem moving the plug away from the seat, thereby opening the valve
- The valve positioner cam is connected to the stem at point x

There are many variations of the control valve, depending on the type of process, the process requirements and the industry. Figure 2 represents the valve flow characteristic which is the relationship between the flow rate of process through a valve and the travel of the stem as the plug is moved from the fully closed position to the fully open position for an ATO valve. This relationship is measured when a constant pressure drop across the valve is achieved, which rarely exists. There are three variations of flow characteristics, viz. linear, equal percentage, and quick opening. The physical dynamics of the internals of the valve body determines these characteristics. The linear characteristic type is commonly used since most control algorithms are linear.

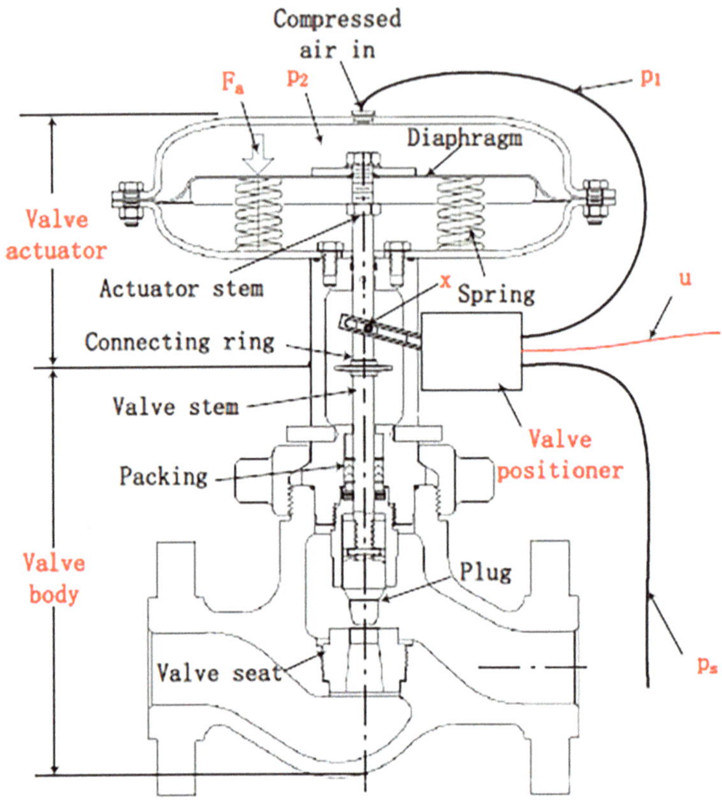

Fig. 1 Cross section of pneumatic ATO control valve [5]

However, in nonlinear control algorithms, especially as we move into the space of AI, equal percentage and quick opening will more often be used.

2.4 The Valve Positioner

Although a control valve can function on its own, the valve positioner is very often required, as it serves certain vital functions, which is the first level of adapting the control valve to optimise control. A valve positioner regulates the specific position of the valve, by serving as feedback from the valve stem to ensure that the position of the valve is proportionate to the control signal from the controller. This is typically achieved by the positioner generating a pneumatic signal of 20–100 kPa to the valve, correspond to the valve's desired position, from fully closed to fully open, for ATO valve. The positioner is located between the controller and the valve actuator, to regulate the air pressure in the actuator for a specific position of the valve. The positioner serves the following critical functions:

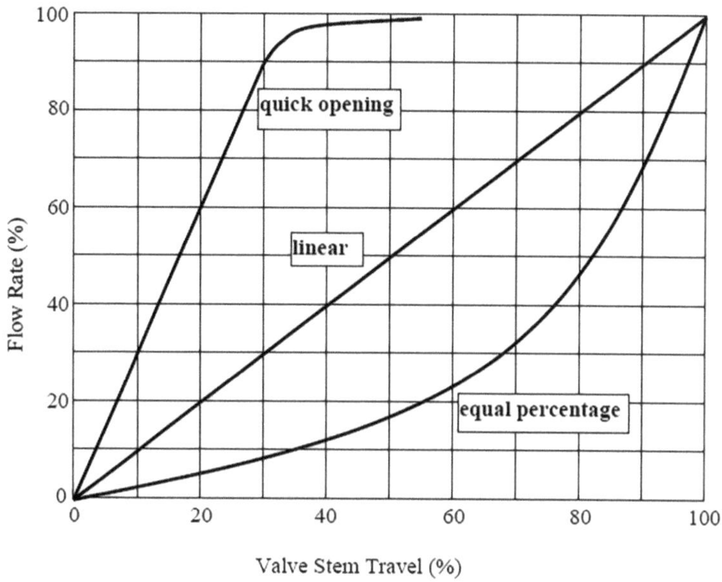

Fig. 2 Valve flow characteristics [4]

- Counteracts external factors like friction, process pressure fluctuations, and metal fatigue, to position the valve to the required position and to regulate the process more accurately
- Reverses the action of a control valve due to the direct (input directly proportional to output) and reverse (input inversely proportional to output) actions, e.g. changes an ATC to ATO valve by inverting the input signal to the positioner when it is generated to the valve
- Speeds up the response time of the valve when the pneumatic input signal varies due to the nature of time lags for volume of air pressure to increase or decrease for the valve to change position
- Incorporates split range control where two or more valves can share the common pneumatic 20–100 kPa signal but the valves operate within certain bands of this signal, for their full travel operating ranges

2.5 The Control Valve as an Instrument in the Control Loop

The four main components of a process control loop are represented in Fig. 3. The control valve, the FCE, is required to regulate the process variable to the desired value as specified in the control element.

Whilst it is important for the measuring instrument to transmit an accurate signal to the controller, it is equally important for the control valve to adjust the process accurately. The main objective of the control loop is to maintain a specified process

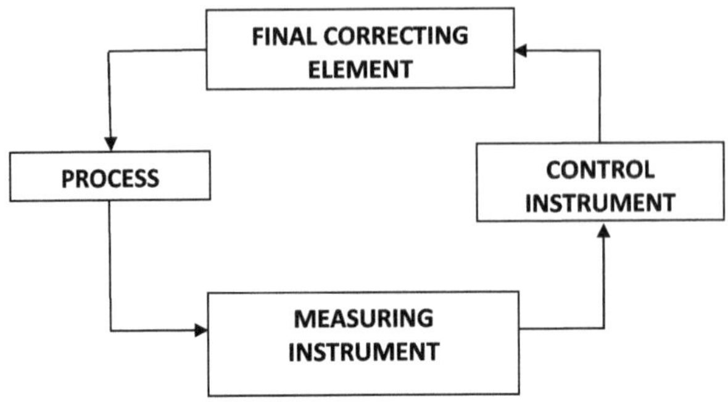

Fig. 3 Components of a process automation and control system [6]

condition, e.g. level, pressure, and flow. The sensor generates a transmission signal to the control instrument depending on the status of the process. Depending on the set point of the controller, a transmission signal is generated to the control valve. The objective is to maintain a zero percent error signal between the process variable and set point. If the error signal increases or decreases from the set point, a corrective signal to the valve results in the physical process varying, until the process is in line with the set point [6].

3 The Control Valve Using AI and ML Technology

Traditional control valves regulate the process medium upon receiving control transmission signals to maintain a specific process variable. AI and ML technology converts control valves from reactive instruments to intelligent instruments. AI algorithms enable control valves to analyse real-time data, commanding them to maintain optimal performance, autonomously. ML algorithms enable control valves to learn from historical real-time data to predict anomalies and plan maintenance schedules. Due to its mechanical operating design, the integration of AI and ML into control valve technology requires sensory instruments to detect specific physical parameters of the valve. These instruments will generate input transmission signals to AI and ML algorithms for corrective transmission to the valve, to physically adjust it to the desired position. The combination of AI and ML with specific sensory instruments converts the traditional valve into an intelligent valve with benefits like improved response times resulting in optimised process control and instrument reliability. Some of the sensory instruments are discussed below.

3.1 Linear Encoder

A linear encoder can accurately measure the control valve stem's position as it moves within the actuator's linear path, when the valve is opening and closing. The valve stem must be in perfect working order, for the control valve to function optimally. The encoder generates a pulsed transmission signal in real time, providing feedback on movement, position, and the direction of movement. This high-resolution data signal is provided as input to the control algorithms in the AI and ML architecture, to analyse the behaviour of the valve in real time. AI algorithms can be used to detect abnormalities in the valve, like wear and misalignment. ML algorithms can analyse this data to recognise patterns to enhance control and develop maintenance schedules [7].

3.2 Rotary Encoder

A rotary encoder can be mounted directly on the valve stem, in the vicinity of the valve body. The valve body must be in perfect working order, for the control valve to function optimally. The encoder measures the exact angle of rotation of movement where the process medium is transported through, converting this rotational motion into a digital transmission signal. In AI-driven systems, high-resolution signal serves as feedback on the state of the internals of the lave body. This is the critical part of the valve where it mounts onto the process line. ML algorithms provide a learning platform to archive historical data, in this case position, wear, performance anomalies, etc. Over a period, the real-time data from the AI system can be used to optimise various operating conditions, improving operating characteristics of the valve. Irregular data patterns in the rotary encoder can serve as alarm alert conditions, to assist with operations of the process.

3.3 Pressure Sensor

A pressure sensor on a process control valve is vital to ensure that there are no leaks. The air pressure signal to the valve must be equivalent to the converted electronic signal from the controller, for the control valve to function optimally. By continuously measuring the pressure on the valve actuator, where the air signal is applied onto the diaphragm, the sensor provides real-time data on the status of the pneumatic control signal. The AI algorithm can analyse this pressure data anomalies, like pressure drops or surges. These abnormal pressure conditions, although not directly related the mechanics of the valve, can be an indication of challenges associated with process leakages, blockages, or wear in the valve mechanism, resulting in an incorrect valve position. ML models can be trained on the archived pressure data patterns to learn on enabling predictive measures.

3.4 Load Cell

A load cell can be used to measure the force exerted by the spring in real time, converting the mechanical force of the spring into an electrical signal, between the diaphragm and the bottom of the valve actuator. The spring tension must be in perfect working order, for the control valve to function optimally. Since the spring tension acts against the air signal, maintaining a specific valve position on an ongoing basis, the properties of the spring metal deteriorate, thereby infringing on the operation of the valve. The signal data from the load cell can be integrated into an AI and ML algorithm to identify any possible malfunctioning of the spring. The algorithm can determine corrective measures like replacement of spring, early detection of spring failure, or the need to readjust the spring tension. AI models can store the data reflecting the normal operating range of the spring tension, resulting in a self-adjusting spring tension without removing the control valve.

3.5 Vibration Sensor

A vibration sensor can be installed on a control valve to measure the state of the diaphragm. The diaphragm is the most delicate component in the valve, separating the air pressure that is trapped on the upper half of the valve bonnet and the spring on the bottom half of the bonnet. Being rubber in its form, it is difficult to establish any deterioration of the diaphragm. The vibration sensor can detect variations in vibration patterns, an indication of deterioration or weakness of the diaphragm. Integrating the data signal with AI and ML algorithms, the sensor provides real-time data on the state of the diaphragm. The diaphragm in the valve bonnet must be in perfect working order, for the control valve to function optimally. ML models analyse this data, compares them to archived data to detect any deviation from the normal condition it should be in, thereby minimising downtime and extending the operational life of the valve.

3.6 Strain Gauge

A strain gauge can be installed on various parts of the valve to measure and monitor metal fatigue on a process control valve. These parts will be primarily the stem and the internals of the valve body. Since the stem acts between the diaphragm and the plug, it is subjected to metal fatigue. Similarly, the process medium flows through the valve body and there is continuous movement with this space, it is subjected to metal fatigue. Strain gauges measure the metal stress (or distortion) detected by the metal that the valve is made up of, to provide real-time data on the valve's structural status. The mechanical/metal properties of the valve must remain within the prescribed limits, for the control valve to function optimally. When installed in critical locations on the valve, these sensors detect the smallest variation in shape, an indication of metal fatigue. These distortions, when measured, are integrated with AI and ML systems, to identify any possible wear against archived data of normal patterns, providing maintenance alerts and generating predictive maintenance schedules, thereby reducing the hazard of valve failure, to ensure reliable and safe process conditions.

4 Machine Learning Techniques on Control Valves

ML results in accurate regulation of the process variable by the control valve, thereby optimising process control. ML is a technique where archived data of specific patterns are accessed, to identify deviations from these patterns and perform certain functions like predicting maintenance requirements, detecting anomalies in the valves, reducing production downtime, and enhancing process reliability, in the interest of improving product quality and safety standards. ML techniques applicable control valve management is discussed below [8].

4.1 Supervised Learning for Predictive Maintenance

Supervised learning is an ML technique whereby archived data is used to train ML models to predict when a valve could fail to function or need repairs. This method uses historical records of valve specifications, sensor specifications, and maintenance to predict imminent conditions facilitating planned intervention in accordance with process requirements, because control valves cannot be removed whilst the plant is running. Predictive maintenance entails measuring and monitoring conditions like process control valve position, flow rate of process, and vibration within the valve, for early detection of deterioration in valve operation, leaks around seals, or blockages in the valve body. Increased vibration in the valve may indicate mechanical defects, whilst irregular flow rate may indicate a blockage or leak in the valve seals.

Supervised learning uses archived data designated as "Okay," "Mechanical failure," and "Leak," membership functions assigned in an algorithm to learn patterns representative of the state of the valve. Common supervised learning algorithms include Support Vector Machines (SVM), random forests, and neural networks. The algorithms obtain complex patterns and relationships between sensor readings and valve condition [9].

4.1.1 Example

In a bottling plant, the filling process is controlled by adjusting the liquid flow into bottles. A trained supervised model uses the archived data from sensors to predict an abnormal valve condition, which triggers the corrective action.

4.2 Unsupervised Learning for Anomaly Detection

The process control valves, in their mechanical form and design, result in their pivotal function in the process. Any failure in their functioning infringes on the overall manufacturing process. Unsupervised learning facilitates automatic detection of abnormal behaviour of the valve in the absence of assigned data. Large volumes of unlabelled sensor data like temperature, pressure, or flow rate are used in unsupervised learning, indicating normal conditions. The fundamental principle is based on the system learning these typical patterns and relationships of the stored data. Any variation from the stored data is representative of an abnormal condition.

Common unsupervised learning systems are the K-means clustering system and autoencoder technique, a type of neural network system. In process control valves, the anomalies can be identified as a blocked valve, degradation in valve condition, or a valve requiring re-calibration, which are detected when the measured data drifts away from stored data. The identification of these anomalies alerts the engineering team to attend to it.

4.2.1 Example

In a batching plant, control valves regulate the flow of various processes. In a K-means clustering algorithm, the ML model identifies groups of data representing normal valve operating conditions which becomes the model. When measured data points deviates from this group of data, it triggers an alarm condition.

4.3 Reinforcement Learning for Adaptive Control

In traditional control methods, the Proportional-Integral-Derivative (PID) controllers work well in stable environments. Furthermore, when a controller is tuned for a process condition, with a specific set point, the process does not function as optimally stable if the set point changes. PID control has limitations in complex processes when nonlinear variables like pH have to be controlled. Reinforcement Learning (RL) offers the solution by enabling an adaptive control approach to enhance valve operation [10, 11].

RL is a subset of ML where a driver learns to make decisions in relationship with the environment. The driver receives feedback in the form of signals from sensors based on the actions, gradually improving the actions to improve the signal against a desired value. Essentially the RL driver determines the valve position to maintain desired process variables, e.g. temperature, despite other varying conditions. An RL-based system learns from historical data and knowledge instead of relying on static tuning as found in traditional controllers. Q-learning and deep Q-networks are RL techniques, depending on the control requirements. The main advantage of RL in valve control is the ability to reduce overshoot and undershoot in the control loop and reduce energy consumption. RL can ensure stable control under varying load conditions or unexpected disturbances.

Whilst the implementation of RL in valve control requires a broad band of training data and extensive processing resources for the data, to ensure stability is a challenge, the advancements in hardware and real-time training make RL-based adaptive control possible [11].

4.3.1 Example

In a chemical processing plant, maintaining the correct pH is critical during reaction stages. An RL model regulates the control valve position dynamically in response to real-time data instead of control settings. The RL algorithm learns the best valve position to maintain the pH under each varying process condition.

4.4 Deep Learning for Complex Pattern Recognition

Deep Learning (DL) is a subset of ML, offering advanced options in composite pattern recognition to further improve the performance of control valves by adjusting to complex patterns in real time. DL models like convolutional neural networks (CNNs) and recurrent neural networks (RNNs) are common in recognising complex and nonlinear relationships within broad bands of datasets. When applied to process control valves, DL can analyse and predict patterns in process and plant behaviour through the various sensors that provide the raw data, to primarily determine action to be taken in relation to the control valve in a loop [12].

A DL model could be trained on archived data specifically from valve sensors, to detect vibrations, metal fatigue, or pressure variations. Any anomalies in these signals will be detected by the DL algorithm as early signs of malfunction of the control valve such as degradation of the spring, blocked valve body, or pressure leaks. DL facilitates adaptive control strategies by continuously adjusting valve position based on process conditions, to adapt to variations in process challenges and ambient conditions [13].

4.4.1 Example

In the boiler plant, Long Short-Term Memory (LSTM) networks analyse successive data from steam control valves, such as temperature, flow, and pressure variations in real time. By identifying patterns like excessive temperature or decreasing flow rate or continuing pressure build-up, the model can predict potential valve anomaly.

4.5 Transfer Learning for Rapid Model Deployment

Transfer Learning (TL) takes a trained model on a specific task and fine-tunes it further, for a specific application. An example, where a ML model is trained on air pressure data that can be adapted for process control valves, by re-training the developed model for a different valve-specific dataset. This pre-trained model retains knowledge of patterns like air pressure fluctuation that are relevant across different types of instruments, including control valves. Fine-tuning the model for valve-specific parameters such as metal fatigue, spring tension, state of diaphragm requires considerably less data than developing the original model [14].

Companies can save on cost and time to gather large amounts of datasets specific to each valve type with the use of TL algorithms, by accessing a generic model for adaption for individual valve types. TL can involve using a CNN or RNN algorithm initially trained on general instrumentation data. The model can be further developed with a smaller dataset for specific valve characteristics. With the advantages of TL comes the challenge in ensuring that the knowledge stored in the pre-trained model is adaptable to various types of valves.

4.5.1 Example

A paint manufacturing company with various types, colours, and grades of paint can use TL to apply a model initially trained on a sheen, blue, interior paint processing line to a processing line of matt, red, exterior paint. By fine-tuning the pre-trained model with minimal additional data, the control valves on the various additive lines can be adapted to maintain required production targets.

5 Benefits of Using AI and ML on Control Valves

AI and ML offer the distinct benefits of predictive maintenance, real-time optimisation, and optimised process conditions, to process control valves. AI-driven algorithms comprise real-time data from sensors monitoring variables like pressure, flow rate, valve position, to continuously analyse and detect anomalies like metal fatigue, pressure variations, facilitating for proactive action before failures occur.

ML-driven algorithms learn the behaviour of control valves under varying operating conditions, enabling more precise regulation of the valve in response to process specifications. This intervention reduces maintenance costs, improves product quality and operational safety, enhances overall system reliability, reduces downtime, and optimises the manufacturing process. Benefits of AI and ML on process control valves are discussed below [15].

5.1 *Enhancing Predictive Maintenance*

Predictive maintenance is an integral component in day-to-day industrial practice, to avoid unplanned machine failures that result in costly downtime and safety risks. Process control valves are continuously adjusting to regulate the process for control of specific variables like flow, pressure, and level. Their continuous operation under various process and ambient conditions makes it essential to function uninterrupted. Traditional maintenance schedules are often faced with challenges of delays due to operational, engineering, or other short falls. AI and ML facilitate real-time monitoring, fault prediction, and scheduling, leading to more effective and consistent operations.

Modern control valves are often equipped with sensors that measure process conditions as well as valve characteristics, generating data. AI and ML algorithms analyse the data, compare them against archived patterns, and identify any variations to detect early signs of anomalies, allowing the system to decide on a corrective measure to be taken. ML algorithms like regression models, decision trees, and RNN are used to effectively identify patterns within complex or nonlinear behaviour.

5.1.1 Benefits of Predictive Maintenance

- Reduced downtime and maintenance costs prevent unwanted part replacements and reduce downtime
- Early fault detection can detect sensitive patterns in data
- Improved safety and reliability ensure that valves operate reliably in the interest of plant and equipment safety
- Optimised control and efficiency facilitate optimal and efficient process control conditions
- Enhanced decision-making facilitates improved decision-making for maintenance and parts replacement

5.2 *Adaptive Control for Improved Efficiency*

Adaptive control is a dynamic model that varies its parameters in real time, with the process environment changes. Conventional control methods, like PID controllers, function within fixed tuning parameters, under certain set point conditions. In adaptive control, the control strategy is based on human reasoning, maintaining process stability under various process conditions. Process control valves are often subjected to process dynamics like fluid viscosity, density, and ambient variations regularly. Such fluctuations require constant adapting to these variations. Adaptive control accesses real-time data from sensors installed on the valves, thereby allowing the control valve to adjust to the process requirements, at the same time facilitating the valve to the process fluid dynamics and ambient changes. An example of adaptive control is where RL comprises an algorithm that learns from past control valve and ambient statistics to influence control [16].

5.2.1 Benefits of Adaptive Control with AI and ML

- Enhanced process stability: Adjustment of valve settings maintains stability in the presence of process variations
- Reduced energy consumption: Enable the control valve to function only when required and allows it to open or close optimally
- Improved responsiveness and precision: Responds instantly to changes, maintaining accurate control of the valve position
- Reduced maintenance and operational costs: Can predict potential process or instrumentation challenges, thereby resulting in proactive maintenance, minimising cost
- Better handling of nonlinear dynamics: Capable of learning and adapting to nonlinear patterns, optimising control further

5.3 *Intelligent Process Optimisation*

Due to the mechanical nature of process control valves, they need to readjust each time the process conditions vary, whilst the physical dynamics of the valve may vary at the same time. AI and ML techniques enable the dynamics of the valve to be compensated for, whilst the process control objectives are maintained. AI algorithms and ML models analyse real-time data from valve sensors to recognise patterns like abnormal vibrations, to optimise the control strategy. ML learning models learn the nonlinear characteristics of the process and valve responses. In this way, the model recognises the relationships between controller input signals and expected output signals, and control valve dynamics through sensors on the valve, to allow the control algorithm to control the process.

5.3.1 Benefits of Intelligent Process Optimisation

- Reduced energy consumption can continuously learn and adapt to updated data, regulating control actions in real time to maintain valve performance under varying process conditions
- Real-time anomaly detection and enhanced safety identify anomalies in control valve behaviour leading to avoiding process instability and preventing dangerous process environment
- Data-driven decision-making facilitates the analysis of data to identify trends, understand failure patterns, and to prevent production and operational bottlenecks

5.4 Fault Detection and Diagnostics

Fault Detection and Diagnostics (FDD) is a critical component of AI and ML when it relates to process control valves because control valves are vulnerable to mechanical wear, cavitation, blockages, and leaks. These factors compromise the control objectives, creating an unstable process. AI- and ML-based FDD systems are advanced solutions to monitor and diagnose various anomalies of the valve, to predict solutions.

5.4.1 Benefits of Fault Detection and Diagnostics

- Enhanced predictive maintenance does not rely on fixed intervals or reactive responses when challenges arise but identifies deviations from normal operating patterns to predict imminent challenges
- Real-time fault detection detects faults in real time, for immediate corrective actions, by using ML models like LSTM networks to analyse sensor data and recognise patterns-related specific faults
- Improved diagnostic accuracy: Classification algorithms like decision NN can analyse data patterns and identify the type of fault, e.g. cavitation and metal fatigue and can distinguish between similar fault conditions
- Reduction of downtime and operational costs: FDD systems detect unforeseen valve failures that can lead to process interruptions, causing production delays and significant financial losses, by identifying faults before they arise
- Continuous learning and adaptability continue to learn and upgrade the fault detection strategies to enhance diagnostic capabilities
- Increased safety and compliance: FDD systems using AI and ML improve overall plant safety to prevent leaks in process lines, explosions, or pollution. They ensure that valves operate in compliance with industry standards and regulations [17]

6 Challenges and Considerations

AI and ML benefits process control valves in many ways, at the same time industry is faced with challenges that require considerations to ensure optimal application. A primary challenge relates to the data being processed. This is followed by integrating AI and ML technology into existing systems. The capital investment in relation to running costs needs careful consideration to establish the return on investment. Security and privacy, when dealing with data and access, make information available but access control is important. These factors are discussed below.

6.1 Data Quality and Availability

High-quality data must be available for training ML models for effective valve regulation. If data quality is compromised, it leads to erroneous predictions. High-resolution data warrants AI algorithms to consistently identify patterns, representative of the real-time transmission signals, to detect anomalies on the physical valve and its performance, and dynamics of the process, to make accurate predictions when required. This data can represent feedback gathered from sensors on the valve, measuring variables like air pressure, metal fatigue, spring tension, and sensors in the plant measuring process variables like level, pH, and density. This data may be in the form of archived breakdown, maintenance and operational records, and is critical for the intelligent system to process for accurate corrective action when required. Unpredictable, inadequate, or unsteady data infringes on model accuracy, thereby leading to errors in positioning the valve correctly.

Methods of data-cleaning processes and sensors that are routinely calibrated are implemented to improve data reliability. This includes the use of only real-time data, for real-time adjustments to take place, as deviations from the norm are detected. IIoT systems enable continuous data collection and sharing, always optimising the control valve position [18].

6.2 Integration with Existing Systems

Very often, whilst existing systems are functioning in the absence of challenges, incorporating any new system creates challenges, especially since the valve is the final component in a loop, to establish the desired process condition. As such, this scenario is created when AI and ML systems need to be integrated into existing systems. A core requirement is the planning stage, to ensure adaptability and compatibility factors are in place, with minimal interruption of plant operations. Since AI and ML architectures require the flow of high-quality data, existing systems will require upgrades on field sensors and communication structures.

Many industrial systems can be modified to accommodate smart sensors with edge computing components, thereby facilitating the processing of data from control valves on a Local Area Network (LAN) and Wide Area Network (WAN), to finally transmit the data to a centralised system for analysis and further processing. Existing systems may comprise systems like Programmable Logic Control (PLC) or Distributed Control Systems (DCS) or Supervisory Control and Data Acquisition (SCADA) that will require a bump-less transition, employing options of using middleware or Application Programming Interfaces (APIs) to allow smooth-flowing data between AI-driven architecture and traditional architecture, both hardware and software. Successful integration will depend on data that is scalable, compatible solutions that complement AI-driven visions, to enable intelligent valve diagnostics without compromising process operations [19].

6.3 Cost and Investment

The capital investment in AI and ML technologies is relatively high, which requires an evaluation of the cost-benefit ratio in medium- to long-term periods, especially since the control valve is the one physical component that determines the final status of plant conditions. Implementing AI and ML technology on process control valves can incur costs that will vary depending on the magnitude and structure of the system. Preliminary expenses include the installation of intelligent sensors and IIoT devices, which collect real-time data from valve and process sensors for machine learning models. The integration of AI and ML systems require dedicated hardware like edge computing devices or cloud infrastructure for the processing of large datasets. Apart from the technology investment, human resource investment is required in either upskilling existing scientists, engineers, technicians, and artisans or employing qualified and skilled personnel. Running costs include data storage and model updates to transition to changing operational conditions.

The above Return on Investment (ROI) can offer benefits like reduced production and machine downtime, real-time predictive maintenance, and optimised plant operations and environment. These benefits can offset the costs, resulting in increased plant and equipment reliability, improved process efficiency and decreased maintenance expenses which can be reduced over time if managed appropriately. For bigger industries with elevated operational and maintenance costs, the generated savings will most certainly outweigh the initial capital investment, leaving the option of implementing AI and ML a valuable long-term strategy.

6.4 *Security and Privacy*

A critical component in AI and ML systems is the protection of sensitive data and safeguarding the cybersecurity of the systems, to facilitate access control and prevent any form of sabotage. These systems are totally reliable on the software data, received in real time from sensors located in and on the control valves. The data is transmitted and received, to and from the control valves, over LAN and WAN infrastructures, making it susceptible to cyber threats. Privacy concerns relate to data collected or transmitted to external vendors as well. Unauthorised access could result in a complete shutdown of the entire factory or critical parts of the factory like the boiler plant. This could lead to equipment failure or safety hazards endangering machines and people, amongst many others.

These risks are managed by securing encryption protocols data pathways, and network access is strictly controlled. Methods of data privacy compliance must be adhered to, for constructive data handling policies, to protect both proprietary information and operational integrity. Cybersecurity measures include regular susceptibility assessments, secure verification and real-time monitoring, to safeguard AI-enabled control valves from developing threats.

7 Future Prospects

AI and ML do not only enable control valves to respond to specific instructions but enables valves to learn from data patterns, adapting to varying process and ambient conditions. In this way, anomalies are predicted preventing unforeseen challenges, thereby safeguarding the control valve status and process, ultimately converting traditional control valves into intelligent, self-optimising FCE. As industries embrace IIoT frameworks, AI-aligned control valves will be an integral component in the control loop, thereby creating fully autonomous systems with devices communicating and optimising processes jointly. Real-time capabilities will develop further, enabling more optimised operations with AI and ML taking process control valves to a higher level of smarter and safer instruments. Some of the future prospects of AI and ML in relation to the process control valve are discussed below [20].

7.1 *Edge Computing*

Sensor data from control valves is normally transmitted to centralised cloud servers for analysis, introducing inactivity of the valve during certain periods of operation, which becomes critical in time-sensitive industrial processes. Edge computing alleviates this by facilitating real-time AI processing directly at the valve location. This allows for instantaneous regulation of the valve based on rapid data analysis. This local processing improves operational resistance, as edge devices can continue functioning independently of network connectivity. This ensures continuous control amidst network disruptions. Edge computing serves the role of improving data security and privacy. This is achieved by not requiring transmitting sensitive data over networks to remote servers, thereby decreasing susceptibility to cyber threats, facilitating control valves to respond quicker, on a secure network [21].

7.2 Explainable AI

Explainable AI (XAI) on process control valves improves data transparency, thereby enhancing factory operations by facilitating production and technical personnel to comprehend complex decision-making processes more easily. Traditional AI models that operate on deep learning techniques often function as "black boxes," creating difficulty in interpreting the reasoning predictions. This complexity creates further challenges to understand the actions applied by an AI-driven control valve, especially where safety is concerned. XAI methods create solution by providing clear, concise intuitions into the model's reasoning. Descriptive data words describe the valve or process variables being measured, e.g. air pressure or process pressure. XAI facilitates safer, more reliable decision-making in process control, improving system performance to meet regulatory standards, whilst maintaining control over industrial processes [22].

7.3 Integration with IIoT

The future of integrating process control valves with IIoT results in an enhancement in the level of intelligence of industrial systems. This enables real-time monitoring of data and autonomous control of devices on a network. IoT-enabled process control valves, apart from communicating with the controller, can communicate with other field equipment sensors and sensors on the valve. This total integration facilitates improved predictive maintenance, a major factor in factory operations. It is achieved by real-time monitoring and analysing data signals received from sensors, tracking performance of the devices and identifying potential risks, to initiate corrective action before a failure arise. When control valves are part of an IoT framework, data falls within a centralised platform, facilitating AI and ML algorithms to analyse the data and train systems to optimise processes effectively and improve operational productivity across entire factory. This configuration also enhances safety measures by through rapid fault detection, thereby making IoT-integrated process control valves a cornerstone of smart factories [17].

7.4 Human–AI Collaboration

AI and ML on process control valves are developed through human–AI collaboration. Hence these systems are based on intelligent decision-making processes. As such, AI provides real-time data, to perform predictive functions for corrective actions, and to continuously make automated adjustments for valve operations as per process requirements. Whilst the control system executes its functions, the human operator interacts with technology with contextual knowledge and problem-solving abilities. AI and ML primarily monitor process and valve status, amongst other variables. At times, human expertise may be required to interpret the system data against unknown variables, e.g. a production variable that changed after system configuration was completed. There are times when there is misalignment between Information Technology (IT) and Operational Technology (OT) that may give rise to unexpected situations or anomalies that AI may not be able to resolve. Furthermore, the human element is an integral component to refine AI models when this misalignment occurs. This collaborative approach influences the roles of both human intelligence and AI, for more robust regulation of process control valves [23].

8 Conclusion

AI and ML technologies have transformed the operational capabilities of traditional control valve systems. They have enabled these systems to become predictive and adaptive, by finding intelligent solutions to optimise process control performance, reduce various operating costs, and enhance safety and reliability of the control valve, personnel, and the environment. AI and ML can predict potential failures and generate respective corrective actions. Analysis of historical data empowers these systems to detect specific patterns that indicate early signs of defects on the control valve or anomalies on the process environment. In this way, predictive maintenance measures are developed to prevent unplanned downtime and to extend the lifespan of the control valve, facilitating reliable operation.

ML algorithms optimise process operations by enabling the control valve to dynamically adapt to varying valve conditions or process conditions. Traditional PID controllers depend on a set of tuning parameters that are limited when process requirements change. AI-driven systems learn from assigned data relevant to the various process requirements or valve specifications, to enhance the position of the valve, the response time for the valve to change position, and the amount of regulation with respect to time, thereby meeting process requirements, in an energy-efficient process environment. AI and ML on control valve systems create the ability to access and apply huge amounts of data from sensors, both from sensors measuring process variables and sensors measuring valve status. AI can simulate complex process conditions, using the stored data, to make predictions on enhancing system design and planning operational approaches.

However, AI and ML systems require high-quality data, specific cybersecurity procedures, and skilled and qualified personnel to manage associated control valve systems. The initial capital investment in implementing such technologies can be substantial but the ROI, in the long-term, benefits industry. AI and ML bridge the gap between traditional systems and intelligent process control technology, setting new standards in process automation and control, creating prospects for further innovation and improvement.

References

1. Bhatia, A.: Control Valve Basics: Sizing and Selection Credit: 4 PDH
2. Oborkhale, L.I., Okoro, C.K., Yerima, O., Onuoha, C.: Prediction of control valve cavitation using machine learning technique. IOSR J. Electr. Electron. Eng. **15**(5), 26–35 (2020). www.iosrjournals.org
3. Ekster, A., Alchakov, V., Meleshin, I., Larionenko, A.: Modeling performance of butterfly valves using machine learning methods. Sustainability (Switzerland). **13**(24), 13545 (2021)
4. Van Vuuren, G.J.: An Introduction to Process Instrumentation. Quad Technologies, 184 p (2010)
5. Tang, L., Fang, L., Wang, J., Shang, Q.: Modeling and identification for pneumatic control valves with stiction. In: IFAC-PapersOnLine, pp. 1244–1249. Elsevier B.V. (2015)
6. P Naidoo. The Control of a Multi-Variable Industrial Process, by Means of Intelligent Technology. 2001
7. Krishna Reddy, C.V.: Prediction of control valve stem position using machine learning with image feature extraction algorithm. J. Crit. Rev. **8**, 534 (2021)
8. Filo, G.: Artificial intelligence methods in hydraulic system design. Energies (Basel). **16**(8), 3320 (2023)
9. Takami, G.: AI-Based Plant Control, vol. 63, English edn. Yokogawa Technical Report (2020)
10. El Koujok, M., Amazouz, M.: Artificial intelligence to improve process control from an energy efficiency perspective: a review
11. Siraskar, R.: Reinforcement learning for control of valves. Mach. Learn. Appl. **4**, 100030 (2021)
12. Sircar, A., Yadav, K., Rayavarapu, K., Bist, N., Oza, H.: Application of machine learning and artificial intelligence in oil and gas industry. In: Petroleum Research, vol. 6, pp. 379–391. KeAi Publishing Communications Ltd. (2021)
13. Kreutmayr, F., Imlauer, M.: Application of machine learning to improve the performance of a pressure-controlled system (2020)
14. Henry, Y.Y.S., Aldrich, C., Zabiri, H.: Control valve stiction detection by use of AlexNet and transfer learning. In: E3S Web of Conferences. EDP Sciences (2021)
15. Jenny, H., Wang, Y., Alonso, E.G., Minguez, R.: Using Artificial Intelligence for Smart Water Management Systems. ADB, Manila (2020). https://www.adb.org/publications/artificial-intelligence-water-supply-ufw
16. Apostolidis, A., Stamoulis, K.P.: An AI-based digital twin case study in the MRO sector. In: IFAC-PapersOnLine, pp. 55–62. Elsevier B.V (2020)
17. Naidoo, P., Sibanda, M.: Building an entrepreneurial culture in electrical engineering education. In: Focus Conference (2023)
18. Al Dosari, F.H.M., Abouellail, S.I.: Artificial intelligence (AI) techniques for intelligent control systems in mechanical engineering. Am. J. Smart Technol. Solut. **2**(2), 55–64 (2023)
19. Kapelan, Z., Weisbord, E., Babovic, V.: Digital Water Artificial Intelligence Solutions for the Water Sector (2020). www.iwa-network.org

20. Naidoo, P., Sibanda, M.: Integrating artificial intelligence in electrical engineering education: a framework for the fourth industrial revolution. In: Focus Conference (2024)
21. Naidoo, P., Sibanda, M.: Emerging trends and future directions of the industrial Internet of Things. In: From Internet of Things to Internet of Intelligence Transactions on Computational Science and Computational Intelligence, pp. 91–110. Springer (2023)
22. Naidoo, P., Sibanda, M.: Embedded supercomputing for edge devices: a comprehensive review. In: From Internet of Things to Internet of Intelligence Transactions on Computational Science and Computational Intelligence, pp. 157–173. Springer (2023)
23. Naidoo, P: Intelligent Control and Tracking of a Solar Parabolic Trough (2006)

Deep Recurrent Neural Network in Machine Translation from English to Kichuka

Tony Munene Gitari, Edna Chebet Too, and Humphrey Kirimi Ireri

1 Introduction

Machine translation (MT) is a part of artificial intelligence (AI) study field that helps computers to understand and work with human languages [1]. Machine translation is applied to many areas such as education, tourism, and trade. Machine translation has the advantage of fast speed and low cost compared to human translators.

Africa's MT in language translation is underdeveloped [2], despite Africa having 16.72% of the global population [3]. The key reason for Africa's low development in machine translation is inaccurate machine translations [4] that utilize classical machine translation approaches [5].

Presently, there exists two technological approaches to MT. The first approach is the classical approach, which has several architectures, including statistical-based MT(SBMT), rule-based MT, and hybrid-based MT of the two machine translation architectures. Classical approaches face challenges such as SBMT classical approach's need for human-based parallel text input to process translations, which is time-consuming. SBMT-based MT is the most widely utilized MT in African languages translation [6].

The second approach for machine translation is the Neural Machine Translation (NMT) approach, which has emerged due to recent advancements in the field. NMT employs artificial neural networks to accomplish translation tasks, with a well-known application being the utilization of Recurrent Neural Networks (RNNs) architectures. RNNs employ loops within neural networks, effectively functioning as a type of memory. Thus, RNNs emulate the human brain, whereby previous

T. M. Gitari (✉) · E. C. Too · H. K. Ireri
Chuka University, Chuka, Kenya
e-mail: echebet@chuka.ac.ke

P. Naidoo, M. Sibanda (eds.), *Measurement, Automation, and Control in Artificial Intelligence and Machine Learning*, Signals and Communication Technology, https://doi.org/10.1007/978-3-031-91802-5_8

words and sentences are stored and used to understand new words and sentences. Furthermore, RNNs show great promise in addressing the challenges of the classical approach such as loss of meaning, ambiguity, labor intensiveness, and word-to-word breakdown. Examples of NMT include long short-term memory (LSTM) machine translators. However, it should be noted that RNNs have the challenge of requiring a large digital language dataset which is not widely available for sub-Saharan languages [7].

In this work, two MT models for English to Kichuka machine translation were developed. Kichuka is an African language. Model one was based on SBMT classical approach, and model two was based on LSTM NMT approach. Both models were trained, validated, and tested using a large digital dataset for Kichuka to English named KED21. KED21 consisted of 5000 number of sentences and 5000 number of words. Both models' translation accuracy was evaluated using the Bilingual Evaluation Understudy (BLEU) score. The BLEU score used data from the test set and compared it to the translated text output from SBMT and LSTM models trained automatically. BLEU score were 0–100, with 50–100 representing very high-quality translation [8]. Results of the BLEU scores were then analyzed for translation accuracy, precision, recall, $F1$ score, and mean weighted average precision. The analyses provided conclusive evidence of the most accurate MT approach to an African language. The results analysis was tabulated using a comparative table, thus addressing the knowledge gap of low MT adoption in Africa due to MT inaccuracy.

2 Methodology

The research site was situated at the main campus of Chuka university, specifically within the computer lab based at the department of computer science.

The research study utilized experimental design, with the goal of determining the highest translation accuracy among the two approaches of MT models.

Model one consisted of an LSTM NMT model carrying out English to Kichuka translation, and the second model was a classical SBMT model MT. A total of two models were developed, trained, validated, and tested, and results are recorded.

The two models were created using the online-based TensorFlow framework. The training, validation, and testing of these models took place on the Google Research Colab platform.

A digitized dataset of Kichuka language was utilized for training, validation, and testing the two models. The data set was from the following verified sources: The Bible Translation & Literacy (E.A.) [9], Chuka University Faculty of Humanities and Social Sciences [10], Kenyatta University Repository [11], Mount Kenya University Institutional Repository [12], and the University of Florida Repository [13]. The dataset was named Kichuka to English Digital Data set (KED21) as shown in Table 1. The KED21 dataset size was 5200 Kichuka words and 5200 Kichuka sentences. The dataset size is optimal for training, validation, and testing of the

Table 1 KED21 digital dataset

Kichuka	English
Ugukinya wega	Have you arrived safely
Thie wega	Travel safely
Ugaca gwetu	Are you coming over
Uri mwega aba	Are you fine
Njeru na nyaga ni acai ma miatu	Njeru and nyaga are beehives craftsmen
Njeri nwa acae muro	Njeri is sharpening a stick
Kwaga nthoni	Lacking decorum
Andikithia	Get someone to write for you
Andikwa	Get employed
Mwandikani	Employer
Guceria wira	Looking for a job
Muruti wira	Employee
Andikirwa	Get someone to write for you
Uhuru auga twikire maski	Uhuru has instructed the wearing of masks
Auge	Let him/her say
Nibu	I accept
Arauka	Dawn
Ria	Eat

proposed statistical-based machine translator model which is data efficient and the stacked long short-term memory translator model, which is data intensive [14]. This dataset size was selected based on several factors:

1. Data efficiency: The statistical-based machine translator model (SBMT) is designed to be data efficient, meaning it can achieve good performance with a relatively small dataset [14, 15].
2. Data intensity: The stacked long short-term memory (LSTM) translator model is data intensive, requiring a substantial amount of data to effectively train the deep learning model [16, 17]. A dataset of 5200 words and sentences is sufficient to provide meaningful patterns for the LSTM model to learn [18].
3. Practical constraints: The size of 5200 words and sentences represents a balance between the availability of digitized Kichuka language data and the computational resources required for training and testing the models [19]. Kichuka is considered a low-resource African language, and the dataset size is large enough to ensure robustness while being manageable within the computational limits [20].

The KED21 dataset preprocessing involved the following procedure: KED21 dataset was saved in Microsoft Excel format called KED21.xslx and therefore was compatible with the TensorFlow deep learning framework.

The data processing for the preprocessed KED21 dataset involved standardizing the dataset by removing spaces, symbols, and numerical data. In addition, the KED21 dataset was divided into 70% of KED21 dataset for prototype training; 20%

Table 2 Luganda to English dataset (LED21)

Luganda	English
Kimenya mateeka okubeera n'emmundu	It is illegal to own a gun
Awa nnyo bazadde be kitiibwa	He is very disrespectful to his parents
Omusajja anywa nnyo omwenge	He takes a lot of alcohol
Twasomeseddwa ku kwerwanako	We were educated about self-defense
Ebiseera bya nkuba	It is a rainy season
Bbaasi eteekeddwa okukanikibwa	The bus should be repaired
Ekitundu kyaguze bbaasi empya	The community bought a new bus
Tukyasonda kugula mmotoka	We are still fundraising to buy a car
Sitambulangako ku lugendo luwanvu	I have never traveled a long distance
Saasondera mukolo guno	I did not contribute to this event
Mutabani wange ajja kusikira ettaka lyange	My son will inherit my land
Essomero libanjibwa nnyo	The school is greatly indebted
Obuwanguzi bujja n'okulemererwa kungi	Success comes with a lot of failures
Kino ekirowoozo kya bizinesi kirungi?	Is this a good business idea?
Buli omu yalowooza nti afudde	Everyone thought he was dead
Olw'ekisa kya Katonda akyali mulamu	It is by God's grace that he is still alive
Kansala yakwatiddwa	The councilor was arrested
Yatonnye olunaku lwonna	It rained the whole day

of KED21 dataset will be used to validate model; 10% will be utilized for BLEU test score. The splitting of the dataset was done in alphabetical order.

The external consistency of the model was examined utilizing the Luganda to English dataset (LED21) shown in Table 2.

The model's architecture was implemented using classical MT of SBMT and NMT using LSTM architecture models. Figures 1 and 2 show the model architectures.

The SBMT model was developed from the IBM SBMT model 2 version 6. The LSTM model was developed from the TensorFlow Keras. TensorFlow Keras is an open-source deep machine learning framework. Both models were trained utilizing the same KED21 dataset and parameters of iterations of 300, epochs of 300, and batch size of 128 as they achieved fitting while consuming optimal computational resources.

The model evaluation of both the Long Short-Term Memory (LSTM) model and the Statistical-Based Machine Translation (SBMT) model utilized the BLEU (Bilingual Evaluation Understudy) score metric. The BLEU score was employed to gauge the precision and quality of the translations produced by each model. BLEU score functions as an automated evaluation metric for machine-translated texts, particularly in the domain of machine translation (MT). Equation (1) shows how BLEU score is determined automatically, and Table 3 shows BLEU score ranking on MT accuracy.

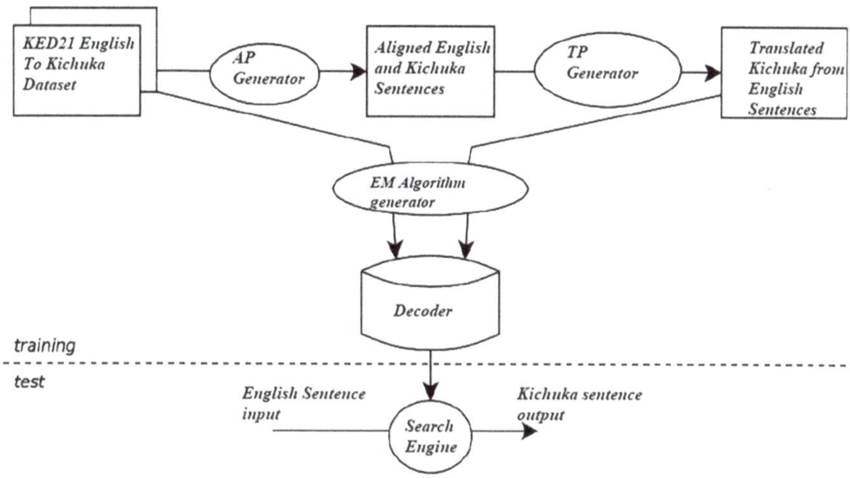

Fig. 1 SBMT architecture for English to Kichuka machine translator

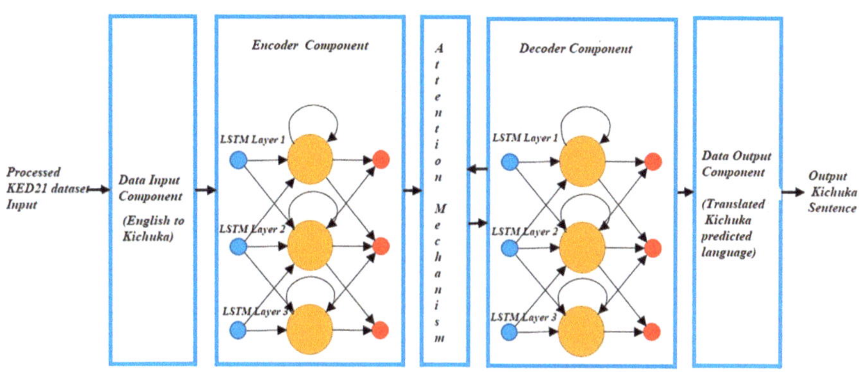

Fig. 2 TensorFlow Keras stack LSTM architecture for English to Kichuka translation

$$\text{Precision}_i = \frac{\sum_{\text{snt}\in\text{cand-corpus}} \sum_{i\in\text{snt}} \min\left(m^i_{\text{cand}}, m^i_{\text{ref}}\right)}{w^i_{\text{t}} = \sum_{\text{snt}'\in\text{cand-corpus}} \sum_{i'\in\text{snt}'} m^{i'}_{\text{cand}}} \tag{1}$$

Equation (1) for BLEU precision [8].

The BLEU precision result was a score value of 0–100.

The results of the BLEU score for the two models were recorded in Microsoft excel format. A comprehensive analysis of the recorded BLEU score results was performed by utilizing a confusion matrix, precision, recall, *F*1 score, and mean average precision (mAP). A comparative table of translation accuracy was used to compare the translation accuracy SBMT against that of LSTM.

Table 3 BLEU score interpretation guide [8]

BLEU score	Machine translation accuracy interpretation
<10	Poor translation
10–29	Low quality translation, grammatical errors
30–49	Understandable to good translations
50–69	Average accuracy translations
70–89	High accuracy, adequate, and fluent translations
90–100	Machine translation accuracy is better than human

```
Epoch 1/300
59/59 [==============================] - 32s 320ms/step - loss: 7.7208 - val_loss: 5.9866
Epoch 2/300
59/59 [==============================] - 15s 258ms/step - loss: 1.6081 - val_loss: 2.1380
Epoch 3/300
59/59 [==============================] - 15s 259ms/step - loss: 1.1165 - val_loss: 1.8607
Epoch 4/300
59/59 [==============================] - 15s 260ms/step - loss: 0.9825 - val_loss: 1.6231
Epoch 5/300
59/59 [==============================] - 16s 264ms/step - loss: 0.8293 - val_loss: 1.3429
Epoch 6/300
59/59 [==============================] - 16s 270ms/step - loss: 0.6973 - val_loss: 1.1845
Epoch 7/300
59/59 [==============================] - 16s 278ms/step - loss: 0.6602 - val_loss: 1.1264
Epoch 8/300
59/59 [==============================] - 17s 290ms/step - loss: 0.6482 - val_loss: 1.0980
Epoch 9/300
59/59 [==============================] - 18s 298ms/step - loss: 0.6411 - val_loss: 1.0785
Epoch 10/300
59/59 [==============================] - 17s 289ms/step - loss: 0.6355 - val_loss: 1.0648
Epoch 11/300
59/59 [==============================] - 16s 279ms/step - loss: 0.6306 - val_loss: 1.0521
Epoch 12/300
59/59 [==============================] - 17s 293ms/step - loss: 0.6259 - val_loss: 1.0417
Epoch 13/300
59/59 [==============================] - 17s 297ms/step - loss: 0.6212 - val_loss: 1.0313
Epoch 14/300
59/59 [==============================] - 17s 289ms/step - loss: 0.6166 - val_loss: 1.0222
```

Fig. 3 Training sample for LSTM English to Kichuka model

3 Results Analysis

This section presents the findings from the training and evaluation of the LSTM model designed to translate English to Kichuka. The analysis focuses on the model's performance, examining the accuracy and efficiency achieved after training with the KED21 dataset, which includes 10,400 Kichuka sentences. The results are presented through various visual aids, such as training process snapshots and model training summaries, to provide a comprehensive understanding of the model's behavior and the effectiveness of the deep learning approach. Figures 3 and 4

```
Layer (type)                  Output Shape           Param #    Connected to
==================================================================================
input_1 (InputLayer)          [(None, None)]         0          []

input_2 (InputLayer)          [(None, None)]         0          []

embedding (Embedding)         (None, None, 256)      1145344    ['input_1[0][0]']

embedding_1 (Embedding)       (None, None, 256)      1864192    ['input_2[0][0]']

lstm (LSTM)                   [(None, 512),          1574912    ['embedding[0][0]']
                               (None, 512),
                               (None, 512)]

lstm_1 (LSTM)                 [(None, None, 512),    1574912    ['embedding_1[0][0]'
                               (None, 512),                      'lstm[0][1]',
                               (None, 512)]                      'lstm[0][2]']

dense (Dense)                 (None, None, 7282)     3735666    ['lstm_1[0][0]']

==================================================================================
Total params: 9,895,026
Trainable params: 9,895,026
Non-trainable params: 0
```

Fig. 4 LSTM for English to Kichuka machine translation model summary

illustrate key aspects of the training process and the overall performance metrics of the model.

3.1 Training Experimentation Results

The LSTM model was trained with 300 epochs. Figure 5 shows the training and validation loss by epochs. The validation loss monitoring was carried out to prevent overfitting and ensure effective training of model the hyperparameter tunings utilized. Figure 6 is a sample of the BLEU score results from training the LSTM MT of English to Kichuka.

The training loss was slightly better than the validation loss for Fig. 5, which inferred that the training dataset was easier to predict than the validation dataset.

The SBMT model training involves iterations of probabilities matching English to Kichuka words. Thus, SBMT training summary based on comparison to corpus translation accuracy. Figure 7 indicates training curve fitting the training data, and Fig. 8 is the SBMT model sample output BLEU score.

For Fig. 7, the training loss suggests that the model performed well on the training dataset. The higher training loss with higher iterations of 0–80 iterations could indicate that the training dataset initial translation probabilities are difficult for the model. This is due to the model developing new translation probabilities for English to Kichuka. However, the curve flattening from 80 iterations and above indicates

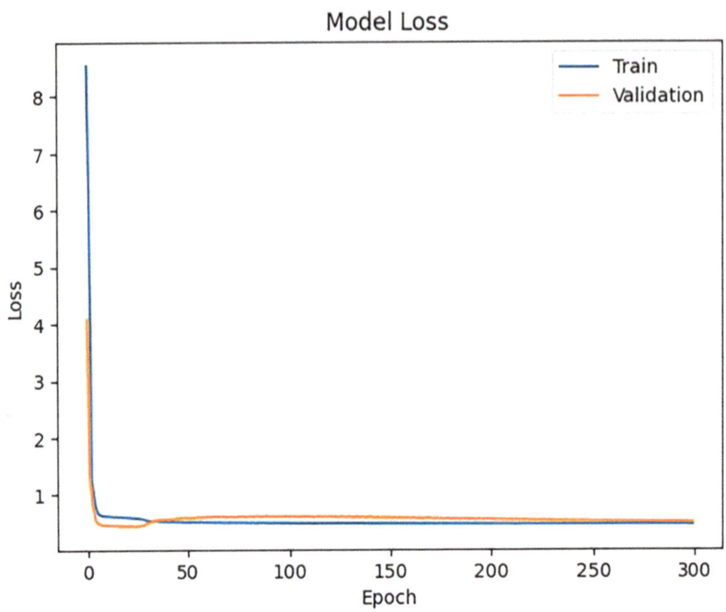

Fig. 5 Training and validation loss by epochs for LSTM MT English to Kichuka model

1	English	kichuka	BLEU Score
2	i wont sing	ntiina	0.951069942
3	hello	muga	1
4	how have you been	watinda atia	0.951069942
5	greetings	mugeni	0.951069942
6	greetings	muugeni	0.951069942
7	i'm at the university	ndi univasiti	1
8	come here	uka baba	0.857143046
9	greetings my teachers	muugeni alimu	0.538173046
10	greetings my panelists	muugeni aa dagit	0.559143046
11	one	imwe	0.951069942
12	two	ciiri	0.951069942
13	three	ithatu	0.951069942
14	four	inya	0.951069942
15	five	itano	0.951069942

Fig. 6 Sample BLEU scores from the training of LSTM MT of English to Kichuka

that the model is training perfectly. This is due to memorization of developed translation probabilities by the model.

The LSTM model accuracy of the training and validation sets was assessed using BLEU score to evaluate how well the model was fitting using the SGD optimizer, softmax activation, and sparse categorical cross-entropy loss functions. The outcomes were depicted graphically in Fig. 9.

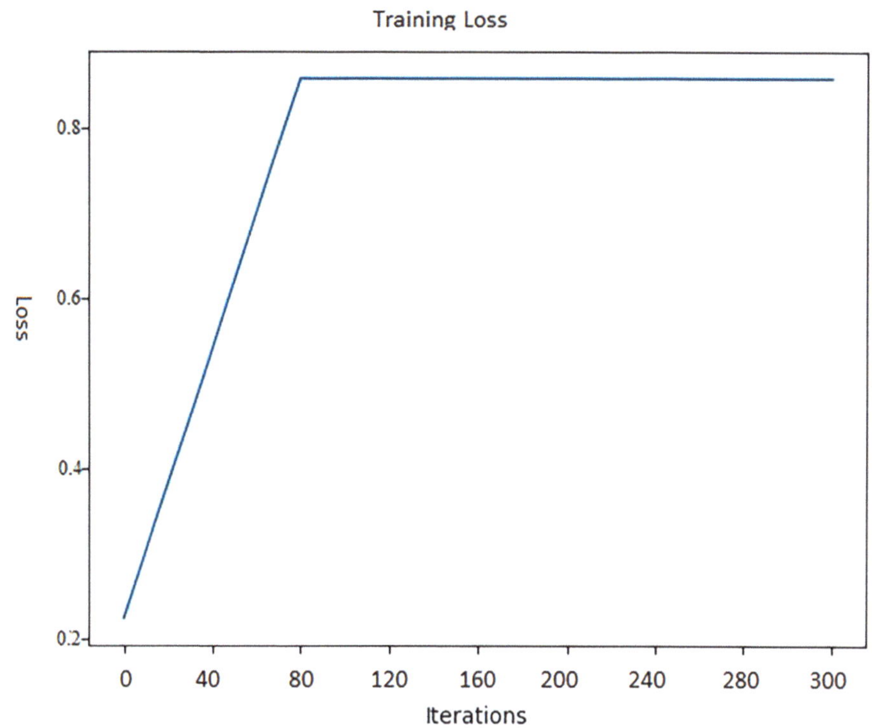

Fig. 7 English to Kichuka training loss curve

In Fig. 9, the validation accuracy was greater than the training accuracy. This inferred that the model generalized well. The spike seen in the training curve was a result of data being loaded in batches.

The SBMT model training and validation accuracy were measured using BLEU score to evaluate the model fitting on the KED21 dataset. The results were visualized in a graph as represented in Fig. 10.

The SBMT model BLEU score was significantly lower than LSTM model BLEU score. The training for SBMT was dropping from 1st epoch to 75th epoch as the model was learning new translation probability rules. From 75th to 300th epoch, the curve became linear, inferring the model adapted to translation probability rules learned in the 1st to 75th epochs.

3.2 *Confusion Matrix*

A confusion matrix provided a summary of the performance of the prediction models. It showed the number of correct and incorrect predictions made by the model for each category. A confusion matrix for each model was generated from the BLEU score results.

Iterations

```
Streaming output truncated to the last 5000 lines.
P(mbere|minutes) = 0.0083026339779l612
P(kurutha|minutes) = 0.01091164979883985
P(wega|minutes) = 4.488761542561182e-05
P(ibinda|minutes) = 0.010603784159331131
P(riria|minutes) = 0.0008872112254538254
P(ria|minutes) = 0.009270127845318958
P(giriru|minutes) = 0.010931700787065563
P(riakinya|minutes) = 0.010931700787065563
P(muntu|minutes) = 0.005786533108165817
P(agaketha|minutes) = 0.003210280505061034
P(kambuge|pcea) = 0.0011919397606854412
P(niwega|pcea) = 0.029844644399639804
P(mutungatiri|pcea) = 0.030261739068255706
P(nikumba|pcea) = 0.031069204322619656
P(ndakinga|pcea) = 0.026714162051642933
P(ciiri|pcea) = 0.007146602357343996
P(niuntu|pcea) = 0.0004964480758798815
P(thiini|pcea) = 0.003311412089070881
P(wa|pcea) = 0.0005255597434601037
P(kanitha|pcea) = 0.002336691347854657
P(pcea|pcea) = 0.031069204322619656
P(ukegua|pcea) = 0.031069204322619656
P(utithiire|pcea) = 0.031069204322619656
```

Fig. 8 SBMT BLEU score output sample

The analysis of Fig. 11 and Table 4, based on precision, recall, and *F*1 scores, reveals the following:

Class 0: The model has relatively lower recall (0.857), indicating that it misses more true instances of this class. The *F*1 score is 0.901, reflecting a balance between precision (0.949) and recall, but with room for improvement in capturing true positives.

Class 1–4: The model performs very well in these classes, with high precision (0.975–0.977) and recall (0.906–0.962). The *F*1 scores range from 0.940 to 0.969, indicating robust and balanced performance across these classes.

Class 5: Precision is lower (0.909), suggesting more false positives, but recall is high (0.984). The *F*1 score remains strong at 0.945, thanks to the high recall.

Overall: The Mean Average Precision (mAP) of 0.961 shows that the model is generally good at ranking relevant instances highly. However, improvements can be made for Class 0 and Class 5, where the model struggles with recall and precision, respectively.

The SBMT English to Kichuka model's performance, as indicated in Fig. 12 and Table 5, reveals the following:

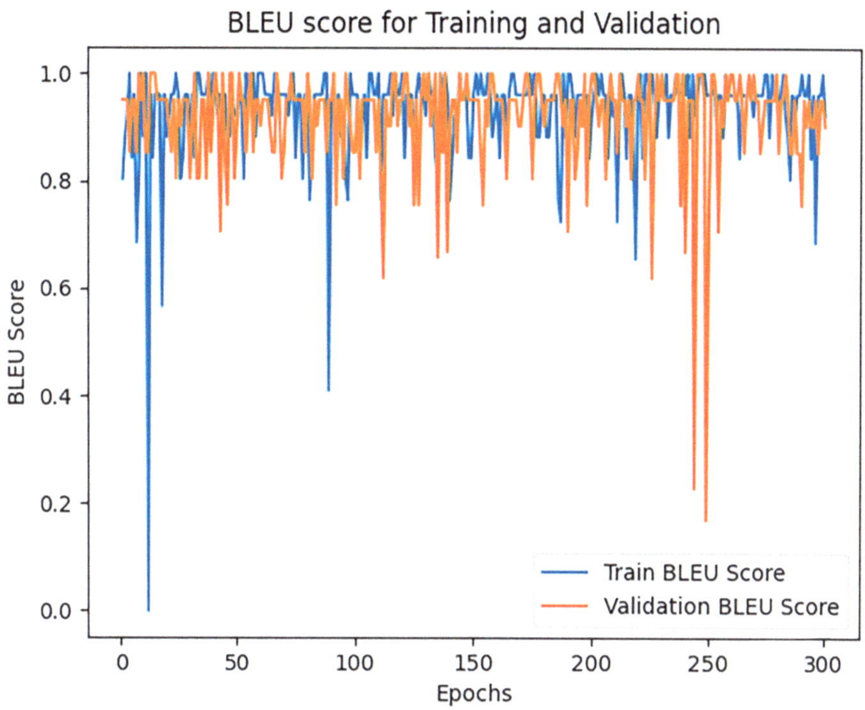

Fig. 9 BLEU score accuracy during training and validation of the English to Kichuka LSTM model

Class 0: The model performs relatively well with a high recall (0.909) but a lower precision (0.473). This imbalance results in an $F1$ score of 0.622, suggesting the model is good at identifying true positives but struggles with false positives.

Class 1: Both precision (0.335) and recall (0.375) are low, leading to a modest $F1$ score of 0.354. This indicates poor performance in correctly identifying instances of this class.

Classes 2–5: The model fails to make accurate predictions, with precision, recall, and $F1$ scores all at 0.000. This reflects the model's inability to correctly identify or predict these classes.

Overall: The Mean Average Precision (mAP) of 0.135 is very low, indicating the model struggles significantly in ranking relevant instances across most classes.

3.3 *Comparative Translation Accuracy Table*

The translation accuracy of the two models was computed. The results were tabulated in a comparative table for analysis, as presented in Table 6.

The LSTM model displayed significant superiority over the SBMT model when evaluated using the KED21 dataset. The performance margin between the LSTM

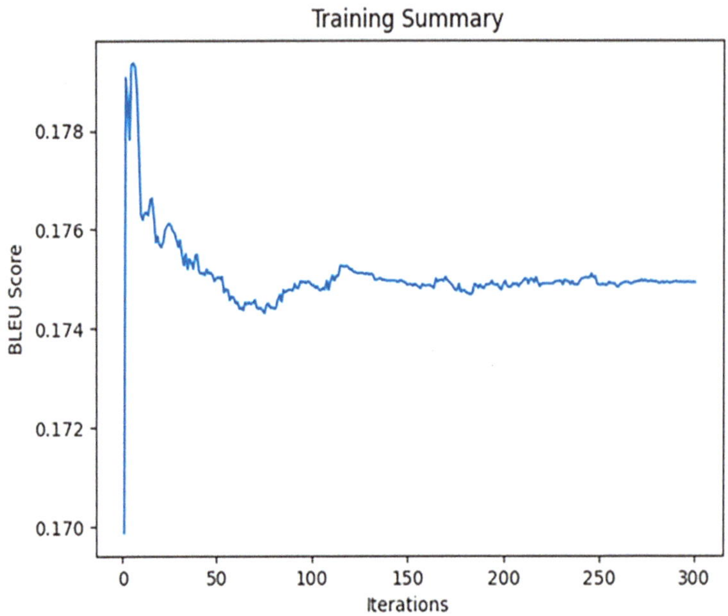

Fig. 10 SBMT English to Kichuka model training accuracy across 300 iterations

and SBMT models amounted to 80%. To ensure an objective assessment, the Mean Average Precision (mAP) values were computed and compared with the Average Precision (AP) values for each model. For the SBMT model, the mAP value was 13% and the AP value was 14%. In contrast, the LSTM model achieved an impressive mAP of 93% for English to Kichuka translation and 92% for Kichuka to English translation. The AP value for the LSTM models was 96%. These findings highlight a substantial difference of 80% in mAP and 79% in AP between the LSTM and SBMT models for English to Kichuka translation. The TL values for the LSTM model were 5% and 4%, respectively, against for SBMT at 84% each, this demonstrates a significant difference of 79% and 80%. The VL value for the LSTM model was 6%. Overall, the experimental results demonstrate the superior performance of the LSTM model in comparison to the SBMT model. Table 6 provides an objective evaluation of the performance of the two models.

3.4 LSTM Machine Translator Comparison to Previous Related Work

The existing MT technologies for African languages work well for textual translations. For example, a robust classical SBMT system for Kiswahili to English translation was developed by the University of Helsinki in 2020 [21]. The BLEU scores

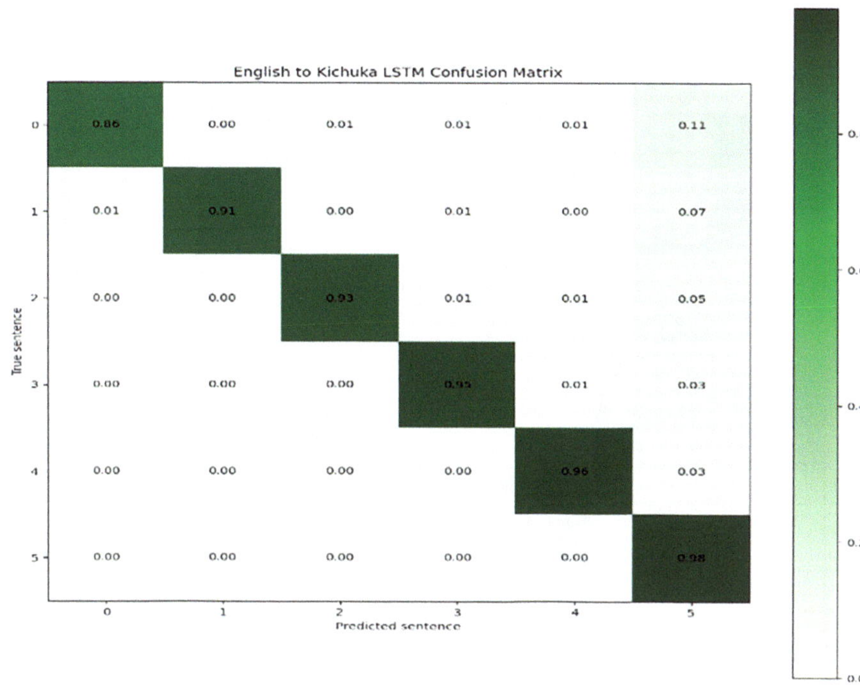

Fig. 11 Confusion matrix for LSTM English to Kichuka model

Table 4 Summary of precision, recall, $F1$ score, and mAP of LSTM English to Kichuka model

Class	Precision	Recall	$F1$ score
Class 0	0.949	0.857	0.901
Class 1	0.977	0.906	0.940
Class 2	0.975	0.926	0.950
Class 3	0.977	0.948	0.962
Class 4	0.976	0.962	0.969
Class 5	0.909	0.984	0.945

Mean average precision (mAP): 0.961

obtained by their SBMT systems for the Swahili-English were 10–15. An RBMT was developed for Tsonga to English translation by the North-West University of South Africa [22]. The BLEU score obtained was 11. An HMT system developed for Somali to English translation by the Johns Hopkins University in 2016 [23]. The BLEU score obtained was 14. A GRU translator for Afrikaans to English was developed by Google Translate and Capetown university [24].

The BLEU score obtained was 71. An LSTM MT was developed by University of Senegal [25]. The BLEU score was 95. It was therefore evident that the LSTM MT performed better than most of the existing SBMT models as shown in Fig. 13.

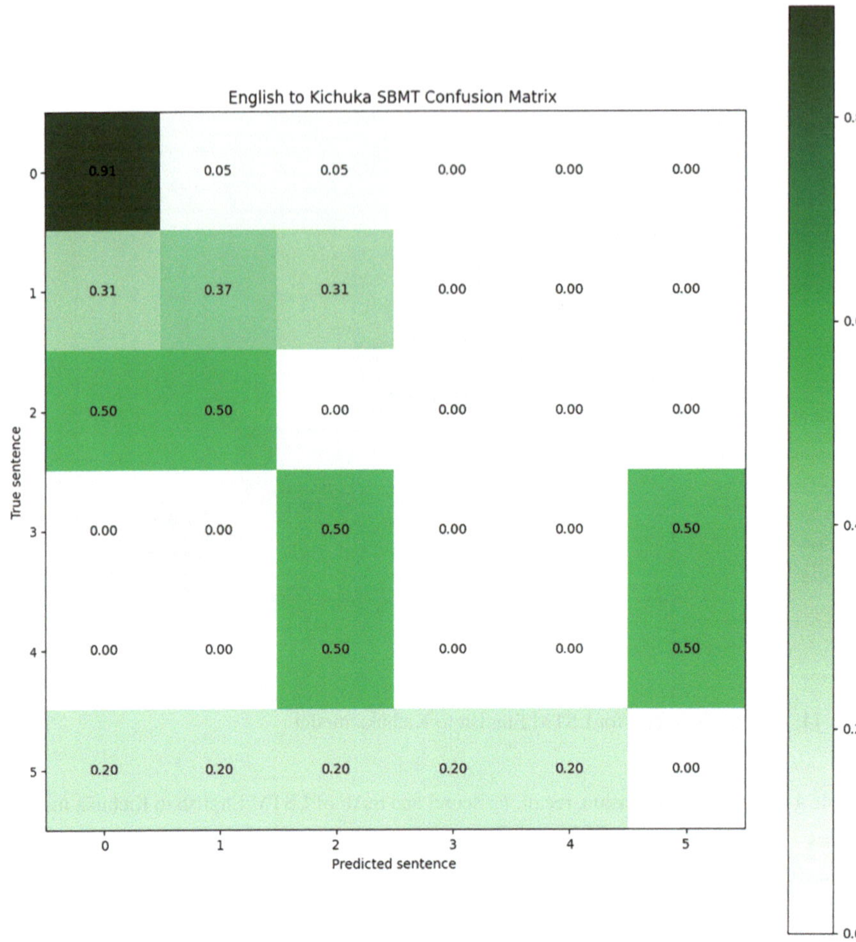

Fig. 12 Confusion matrix for SBMT English to Kichuka model

Table 5 Summary of precision, recall, $F1$ score, and mAP of SBMT English to Kichuka model

Class	Precision	Recall	$F1$ score
Class 0	0.473	0.909	0.622
Class 1	0.335	0.375	0.354
Class 2	0.000	0.000	0.000
Class 3	0.000	0.000	0.000
Class 4	0.000	0.000	0.000
Class 5	0.000	0.000	0.000

Mean average precision (mAP): 0.135

Table 6 Comparative translation accuracy table

Model	Translation accuracy (TA)	Precision	Recall	*F*1 score	mAP (mean average precision)	Training loss	Validation loss
SBMT English to Kichuka	14%	14%	89%	61%	13%	84%	0%
LSTM English to Kichuka	96%	96%	97%	96%	93%	5%	6%

Model	Translation Accuracy	Precision	Recall	F1 Score
LSTM	**96**	**96**	**97**	**95**
SBMT	**14**	**14**	**89**	**60**
(Duh, 2020)	**14**	**-**	**-**	**-**
(The University of Helsinki, 2020)	**15**	**-**	**-**	**-**
(Groenewald and Fourie, 2009)	**11**	**-**	**-**	**-**
(Yonghui and Mike, 2018)	**71**	**-**	**-**	**-**
(Elhadji and Alla, 2020).	**95**	**-**	**-**	**-**

Fig. 13 LSTM against previous models' comparison

4 Conclusion

The study verified that translation model using the long short-term memory network framework (LSTM) exhibited superior performance compared to the statistical-based machine translation (SBMT) model. The LSTM model achieved an impressive translation accuracy of 96%, contrasting starkly with the 14% achieved by the SBMT model. The weighted average precision, recall, and *F*1 score for the LSTM were 96%, 97%, and 95%, respectively, as opposed to the SBMT, which scored 14%, 89%, and 60%.

The research firmly established LSTM as the optimal choice for English to Kichuka translation over SBMT. The LSTM's interpretability stems from its learning through mem-orization, bolstered by efficient parallel computation. It proves particularly suitable for datasets with sentences and words following similar syntax.

Furthermore, the research demonstrated that the model performance could be con-sistently replicated in various settings. This was confirmed by external validation using the LED21 dataset. The LSTM model exhibited a recognition accuracy of 92%, in contrast to the SBMT 13%. The weighted average precision, recall, and *F*1 score of the LSTM were 91%, 95%, and 94%, respectively, compared to the SBMT 13%, 78%, and 61% when tested on the LED21 dataset.

The study's findings were consistently reproducible, highlighting the model's reliability. The remarkable 79% improvement in translation accuracy demonstrated by the LSTMs is especially noteworthy considering the linguistic dissimilarities between Luganda and Kichuka. LSTM stands to become a potent tool with potential to boost trade volumes in Tharaka-Nithi county, stimulate tourism, facilitate more effective Kichuka community outreach initiatives, and contribute to the digitization of English to Kichuka languages.

4.1 Suggestions for Further Study

1. Exploration of transformer models: Future research should investigate the application of transformer-based architectures, such as BERT or GPT, which have shown superior performance in various language tasks, to improve translation accuracy for African languages.
2. Larger and more diverse datasets: Increasing the size and diversity of training datasets by including more sentence structures, contexts, and dialects can enhance the model's robustness and accuracy.
3. Hybrid approaches: Combining classical statistical methods with neural models may offer complementary strengths, potentially leading to higher translation accuracy.
4. Cross-linguistic transfer learning: Research into how transfer learning techniques can be used to leverage high-resource languages to improve MT for low-resource African languages.
5. Real-world application testing: Implementing and testing NMT models in real-world applications across different sectors (e.g., education, healthcare) to identify practical challenges and refine the models based on user feedback.
6. Improvement in preprocessing techniques: Developing better preprocessing and normalization techniques to handle linguistic nuances and variations within African languages.
7. Incorporation of linguistic rules: Integrating linguistic rules and domain-specific knowledge into the NMT models to address specific translation challenges unique to African languages.

4.2 Suggestions for Further Research

1. There is also a need for more research funding into deep learning NLP technologies for African languages translation. Most African languages are translated using the classical MTs of SBMT and HMTs, this is due to the lower costs of computational resources. However, this research and other referenced research papers have shown deep learning to have a much higher translation accuracy.
2. There is also a need for further research funding into the data collection and digitization of African languages such as Kichuka for machine learning datasets.

Larger and diverse digital Kichuka datasets of above 100,000 sentences would achieve superior new sentences predictions.

3. The efficiency of the LSTM frameworks can be further augmented by incorporating supplementary features. For instance, adaptations can be made to enable the inclusion of non-character-based datasets, encompassing numbers and symbols, alongside character-only datasets. This adjustment would facilitate more comprehensive translation outcomes. The incorporation of other deep learning architectures such as sequence to sequence transformers will increase the parallel computation efficiency of the LSTM frameworks and hence lower the costs of training LSTM-based models.
4. The LSTM model can be utilized for training and testing translations based on other Bantu languages such as Kikuyu, Kitharaka, Luganda, etc.

References

1. SAS Insights, "What is Natural Language Processing?" SAS Institute Inc., 2021. [Online]. Available: https://www.sas.com/en_us/insights/analytics/what-is-natural-language-processingnlp.html#:~:text=For%20example%2C%20NLP%20makes%20it. [Accessed: Aug. 19, 2025].
2. Statista Research Department, "Natural Language Processing Market Revenue in Africa 2015–2024," Statista Inc., Hamburg, 2016.
3. United Nations Population Division, "Population, total—Africa," The World Bank, New York, 2019.
4. Knowledge 4 All Foundation Ltd., "Cracking the Language Barrier for a Multilingual Africa," Knowledge 4 All Foundation Ltd., 2020. [Online]. Available: https://www.k4all.org/project/language-dataset-fellowship/. [Accessed: Aug. 19, 2025].
5. A. Buliva, "Machine Natural Language Translation Using Wikipedia as a Parallel Corpus: A Focus on Swahili," Nairobi Metropolitan Area, United States International University Africa, Nairobi, 2019.
6. S. Agbolo, "A Machine Translation Software for African Languages by an African," Kabod Group, 2022. [Online]. Available: https://kabodgroup.com/a-machine-translation-software-for-african-languages-by-an-african. [Accessed: Aug. 19, 2025].
7. T. Tracey, "Language Translation with RNNs," Medium.com, San Francisco, CA, 2019.
8. Google, "Google Cloud Translation Evaluating Models," Google Cloud Translation, 2021. [Online]. Available: https://cloud.google.com/translate/automl/docs/evaluate. [Accessed: Aug. 19, 2025].
9. Bible Translation & Literacy (E.A.), "The Gospel of Luke in Chuka Language," Nairobi, Kenya, 2017.
10. Chuka University Faculty of Humanities and Social Sciences, "Kichuka Language Digital Dataset," Chuka University, Kenya, 2022.
11. M. P. Wambui, "Meaning Loss in the Translated Kimeru Proverbs and Idiomatic Expressions," Master's thesis, Kenyatta University, Nairobi, Kenya, 2009.
12. O. G. Ntiba, "Ushawishi za nyimbo za kampuni wa jamii ya Wachuka katika kaunyi ya Tharaka Nithi, Kenya," Mount Kenya University Press, Meru, 2016.
13. J. Thomas, "African Languages and Translation Studies," Ph.D. dissertation, University of Florida, Gainesville, FL, USA, 1992.
14. T. Cagala and D. Bundesbank, "Improving Data Quality and Closing Data Gaps with Machine Learning," Deutsche Bundesbank, German Securities Holding Statistics, 2019.
15. P. Koehn, Statistical Machine Translation. Cambridge University Press, 2009.

16. D. Britz, A. Goldie, M.-T. Luong, and Q. Le, "Massive exploration of neural machine translation architectures," in Proceedings of the 2017 Conference on Empirical Methods in Natural Language Processing, 2017.
17. S. Hochreiter and J. Schmidhuber, "Long short-term memory," Neural Comput., vol. 9, no. 8, pp. 1735–1780, 1997.
18. X. Chen, C. Liu, D. Yin and J. Tang, "Research challenges and opportunities in natural language processing," Journal of Natural Language Engineering, vol. 24, no. 4, pp. 551–572, 2018.
19. R. Jozefowicz, W. Zaremba, and I. Sutskever, "An empirical exploration of recurrent network architectures," in Proceedings of the 32nd International Conference on Machine Learning (ICML-15), 2015.
20. G. Nakweya, "African Languages Remain Low Resource in Digital Era," University World News, 2021.
21. The University of Helsinki, "The University of Helsinki Starts a New Africa Program and a Strategic Partnership with the University of Nairobi," University of Helsinki, 2020. [Online]. Available: https://www.helsinki.fi/en/news/highereducationpolicy/university-helsinki-starts-new-africa-program-and-strategicpartnership-university-nairobi. [Accessed: Aug. 19, 2025].
22. "LibGuides: Languages: African Languages," [Online]. Available: https://libguides.nwu.ac.za/languages/african-languages. [Accessed: Aug. 19, 2025].
23. K. Duh, "Benchmarking Neural and Statistical Machine Translation on Low-Resource African Languages," ACL Anthology, 2020. [Online]. Available: https://aclanthology.org/2020.lrec-1.325/. [Accessed: Aug. 19, 2025].
24. W. Yonghui, M. Mike, and M. Klaus, "Google's neural machine translation system: bridging the gap between human and machine translation," arXiv:1809.08144, 2018. [Online]. Available: http://arxiv.org/abs/1809.08144. [Accessed: Aug. 10, 2024].
25. M. Elhadji and L. Alla, "Using LSTM to Translate French to Senegalese Local Languages: Wolof as a Case Study," 2020.

Development of a Networked Plug with a Consumer-End Electricity Smart Metering System

Mathias Shimwetheleni, Janvier Kamanzi, and Marco Adonis

1 Introduction

In an era of escalating electricity costs, the need for households to monitor their energy consumption has become more urgent than ever. Real-time tracking of energy usage enables households to indentify opportunities for cost-saving and well-informed decisions regarding their energy consumption. However, accessing this critical data remains a considerable challenge for most households.

Conventional utility bills provide only limited insight into energy consumption and are issued with delays, making real-time monitoring unattainable for consumers. This has created an imperative for innovative solutions that offer direct access to consumption data for individual appliances while also facilitating remote on-off control. The demand for such solutions stems from the growing recognition that consumers require a more responsive and insightful approach to energy management in the face of rising electricity costs.

2 Literature Review

The demand in home automation and advances in Internet of Things (IoT) have triggered the deployment of Smart Plugs (SPs) [1, 2]. The latter have experienced a significant popularity with an estimated growth market of \$0.71 billion in 2016 to \$2.59 billion in 2021 [3], of which the market value is expected to have grown by

M. Shimwetheleni (✉) · J. Kamanzi · M. Adonis
Department of Electrical, Electronic, and Computer Engineering, Cape Peninsula University of Technology, Bellville Campus, Cape Town, South Africa
e-mail: shimwethelenim@cput.ac.za; kamanzij@cput.ac.za; adonism@cput.ac.za

P. Naidoo, M. Sibanda (eds.), *Measurement, Automation, and Control in Artificial Intelligence and Machine Learning*, Signals and Communication Technology, https://doi.org/10.1007/978-3-031-91802-5_9

the present year of this paper. To ensure a comprehensive understanding of the problem at hand, a thorough literature review was conducted. This involved an extensive analysis of previous research and work done by experts in the field and specific features involved in the development of smart plugs.

2.1 Communication Protocol

Different communication technologies are used to provide an interface between the hardware components and the software, and they include Bluetooth, Wireless Fidelity (Wi-Fi), Z-wave, and Zigbee [4, 5]. Many studies on smart plugs focused on Wi-Fi, Bluetooth, and ZigBee technologies as their means of communication protocols. However, a study carried out by [6] identified that these wireless networks had a short range and increase the cost of installation when repeaters become necessary to boost the signal coverage [7]. The authors also identified that in some areas of the house such as garages the signal was weak, which led to difficulties in controlling equipment installed in those areas. In addition, they added that these networks had a high energy consumption which led them to explore Low-Power Wide Area Networks (LPWANs) which included Long Range WAN (LoRaWAN) to address coverage issues and high energy consumption when trying to cover all areas of the house. With each LoRaWAN gateway able to manage about 1000 connected devices, and with a range of up to 5 km, they could greatly reduce the energy consumption of wireless devices [8].

2.2 Security

Security issues reported on wireless technologies including ZigBee, Bluetooth, and Wi-Fi relating to cyberattack vulnerability recommend the use of SPs in safest network possible [9, 10].

Energy usage data collected by these devices can reveal residents' habits when they are away from home, creating a potential security threat. Researchers like Lee and Yang have demonstrated the extent of vulnerability of the likes of Kasa HS100 and Belkin Wemo SPs. Through simulations, the researchers expose security issues such as Wi-Fi sniffing [11]. Additionally [12], identified security issues with the Edimax SP-2101W smart plug, conducting four different attacks, including device scanning, brute force, spoofing, and firmware, which led to the acquisition of user authentication credentials, enabling malicious remote operations and malware installations.

On the environment perspective, nations are working to decarbonizing the world or getting a net zero state of carbon emissions in the atmosphere which leads to global warming [13, 14]. Hence energy efficiency has become an important topic to research with the objective of reducing the production cost of the power stations

[15]. Smart grids or electrical power systems, therefore, make use of IoT devices to monitor the energy use of their respective consumers [16].

When integrating these IoT devices into an electricity grid, a network security challenge should be addressed. A cyberattack on the grid could have financial costs, economic repercussions, even loss of life for those relying on uninterrupted power to live [17]. Denial-of-service, man-in-the-middle, and phishing are major cyberattacks targeting IoT devices in the electricity sector when the communication is not encrypted [18].

2.3 *Energy Demand*

Peak energy demands have high constraints on the power system, leading to large generation capacity needed and eventually increasing electricity production costs [18, 19]. For power stations to remain running profitably means that units cost per kWh needs to be strategically laid out to benefit both consumers and the electricity suppliers. Home Energy Management Systems (HEMS) has an objective to decrease the peak demand of households by controlling power loads flexibly to meet households' priorities [20]. On the demand side, Turkey uses the Time of Use (ToU) tariff system to charge consumers different rates at different times of the day. Their time is divided into the on-peak period between 17:00 and 22:00, the off-peak period between 22:00 and 06:00, and the mid-peak period between 06:00 and 17:00 reflecting the level of energy demand on the power station. This affords cheaper electricity prices during off-peak periods, and by notifying the user through HEMS automated system, a consumer can shift their higher power consumption loads from on-peaks to off-peak periods, hence saving money [21]. Their study further reiterated that by using smart plugs incorporated in their HEMS technologies on particular houses in Turkey for a month, a reduction of up to 33% in peak demand on the power system was achieved.

2.4 *Accuracy*

As with most, if not all smart plugs have functions of measuring load power, a few studies have paid attention to the accuracy of the power and energy measured by these devices. A study on smart plugs from different commercial manufacturers focusing on the meter's accuracy with fast changing load levels as often happens with power electronics devices revealed non-negligible inaccuracies levels ranging between 54% and 100% at switching frequencies ranging between 0.01 and 2 Hz with a reporting delay that can go as high as 20 s. However, data collected on long-duration steady-state loads, which are mainly resistive through heating elements or inductive are exempted from significant measuring errors [3].

As a result, the authors in their study indicate that the use of smart plugs for power research and control applications, often requested by the National Department of Energy in analyzing their energy consumption to balance the supply and demand and to allow users to reduce their carbon footprints, should be viewed with caution and validated before committing to use these devices in their studies.

2.5 Affordance

A study suggests that Smart plugs (SPs) can serve as an energy-saving solution through functions such as standby power cut-off, remote monitoring and control of power consumption [22]. However, energy savings can only be experienced if users adopts SPs, which may be hindered by difficulties in their operation [22]. The study emphasizes the importance of making smart plugs more user-friendly by helping people understand their capabilities and encouraging regular use. It also found that many users did not understand power consumption values expressed in kilo-Watthours (kWh) and preferred information to be presented as electricity costs, which they found more convenient. Another recommended improvement is a compact plug design to prevent interference with adjacent outlets or power strips when the SP is in use [22].

In this research project, a smart plug was studied, designed, and implemented based on Wi-Fi technology. The ESP32 microcontroller was configured as a Wi-Fi access point to connect to the same network as the smartphone. A web server was then set up on the ESP32 once connected to the Wi-Fi network to handle Hypertext Transfer Protocol (HTTP) requests from the phone application. Variable loads, primarily power electronic devices, along with the steady-state loads, were tested to verify the accuracy of the measured data, The design of the plug was kept as simple as possible to maintain ease of use and affordability.

2.6 Project Aim

With current metering systems, customers face the challenge of tracking the energy usage and controlling it in real-time, regardless of distance. This project aim to address the challenge of managing electricity costs by providing a practical solution for households to access their consumption data and remotely control appliances.

The key objectives are: real-time energy monitoring, cost-saving, user friendliness, remote controllability, and affordability. To achieve these objectives, the proposed design must comply with engineering standards in terms of data protection and power quality measurement. The General Data Protection Regulation (GDPR) was consulted to ensure the user information is safeguarded and that the intergration of IoT in this project aligns with the privacy laws [23]. For metering requirements, relevant sections of the International Electrotechnical Commission (IEC) standards guided the development of the networked smart plug [24]. The IEC 62053 was referenced for metering accuracy and safety while the IEC 61010 informed the design for safe operation in various environments [24].

3 System Design

Embarking on the heart of this project, the focus shifts to the intricate design and operation of the smart plug solution. From the core components driving this innovation to the technical schematics guiding its functionality, the hardware and software designs and their interaction formed the main phases. Completion of these phases extends to the user interaction and functionality of the developed prototype, providing a comprehensive understanding of its operational dynamics.

3.1 Conceptual Design

The project's conceptual design consists of a Wi-Fi connectivity, a smartphone, a controller which is an ESP32 microcontroller, a smart plug, appliances, and a server. Figure 1 shows the interaction between these components.

3.2 Major Components

Understanding the major components required in the system development is crucial. The component list is as follows:

3.2.1 ESP32 Microcontroller

The ESP32 serves as the central microcontroller and backbone of the smart plug system. Its low-power consumption and built-in Wi-Fi and Bluetooth capabilities make it an ideal choice for this project. The selection of the ESP32 microcontroller is primarily due to its intergrated Wi-Fi module, essential for the smart plug's communication. The 34 General Purpose Input/Output (GPIO) pins provide flexibility for connecting and managing multiple components effectively. This feature is particularly useful for interfacing with the current and voltage sensors, utilizing the 12-bit Analogue-to-Digital (ADC), which offers more precise measurements compared to the ESP8266.

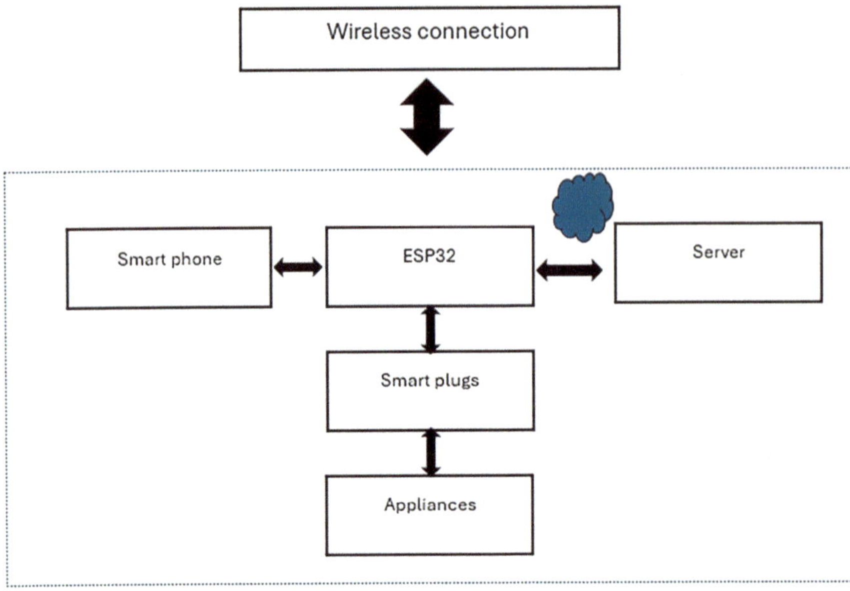

Fig. 1 Block diagram for the conceptual design. The ESP32 microcontroller acts as a central hub, wirelessly connected to a smartphone and a remote server. It communicates with smart plugs, which in turn control house appliances. The entire system operates over a wireless network, enabling remote monitoring and control of devices

Fig. 2 ESP32 Devkit DOIT V1

The dual-core processor, consisting of two Xtensa 32-bit LX6 microprocessors, provides the processing speed required for real time data handling without delays. Although this can result in higher power consumption, the ultra-low-power coprocessor compensates by managing ADC readings and computations during deep sleep mode, thereby conserving energy.

Compatibility with the Arduino programming environment facilitates ease of development by allowing the use of familiar tools and methods. Although debugging can present challenges, the availability of a detailed datasheet aids in effective troubleshooting. The ESP32 is shown in Fig. 2.

3.2.2 Hi-Link 5 V DC Power Supply

The Hi-Link 5 V DC power supply is a fundamental component within the smart plug system, providing the necessary electrical energy to support the operation of the entire setup. This power supply module converts the standard 110–250 V AC voltage from a household outlet into a stable 5 V DC output, which is suitable for powering microcontrollers, sensors, and other components within the smart plug.

A key advantages of the Hi-Link power supply is its compact and reliable design. It incorporates features such as over-voltage protection, short-circuit protection, and high energy efficiency [25]. These features ensure stable and safe power delivery to the system reducing the risk of electrical faults and damage to the connected devices. Figure 3 shows the Hi-link 5 V DC power supply.

The Hi-Link 5 V DC power supply plays a critical role in powering the ESP32 microcontroller, voltage and current sensors, and other electronic components. Its efficiency and safety features are essential for the reliable and continuous operation of our smart plug.

3.2.3 2-Channel Relay Module

The 2 channel relay module is designed to handle both low-voltage control signals from the microcontroller and higher voltage circuits connected to household appliances. Figure 4 shows the 2 channel relay module that was used in our project.

Fig. 3 Hi-Link 5 V DC power supply

Fig. 4 2 channel relay module

Fig. 5 ZMPT101B voltage sensor

3.2.4 ZMPT101B Voltage Sensor

The ZMPT101B voltage sensor is a precision component based on the ZMPT101B voltage transformer. It is specifically designed for the accurate measurement of AC voltage when integrated with an Arduino or ESP32 microcontroller. This sensor is capable of effectively measuring AC voltage up to 250 V and provides an adjustable analogue output (Fig. 5).

More specifications of the ZMPT101B voltage sensor include the following characteristics: Lightweight with on-board micro-precision voltage transformer, high-precision on-board op-amp circuit, operating temperature: 40 to +70 °C, and supply DC voltage of 5–30 V [26].

3.2.5 SCT013 Current Sensor

The SCT013 current sensor is a key component in the smart plug system, designed for measuring electrical current. The following specifications were consindered: input currents of 10 A max, output signal of 0–1 V DC, non-linearity of 2–3%, and a built-in sampling resistance of 62 Ω. Figure 6 shows the SCT013 current sensor [26].

3.2.6 Switches

Switches play a crucial role in controlling the flow of electrical power to connected devices while ensuring user-friendly operation. The inclusion of electronic switches allows remote control through the app, complementing the physical manual switches on the device.

3.2.7 Plug Socket

The plug socket is the physical interface that connects the smart plug to the electrical outlet. Figure 7 illustrates the manual switches and plug sockets used in the project.

Fig. 6 SCT013 current sensor

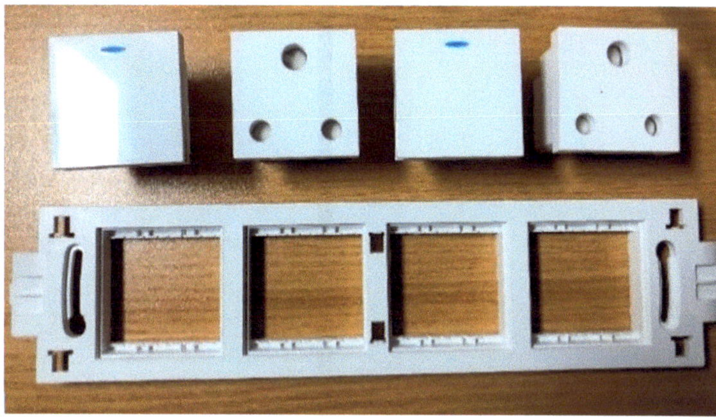

Fig. 7 Physical switches and plug sockets

In this project, switches and plug sockets are essential components that enable users to control and power their devices through the smart plug. Whether via manual interaction or remote control through the smartphone application, these components ensure a seamless, user-friendly experience while optimizing energy management.

3.2.8 Smartphone

The smartphone's wireless connectivity, such as Wi-Fi or cellular data, enables it to communicate with the smart plug's microcontroller. This connection enables users to control their devices from virtually anywhere, providing convenience, remote monitoring, and energy management capabilities. Figure 8 depicts the smartphone interface used used in the project, displaying data recorded according to the parameters of interest and the energy cost.

3.3 Smart Plug Schematic Circuit and Simulation

Next, the technical underpinnings of the system are explored through a detailed exploration of the schematic circuit. This visual map serves as the blueprint for how our components interact. Figure 9 illustrates the schematic circuit of the project, which was designed using Fritzing software.

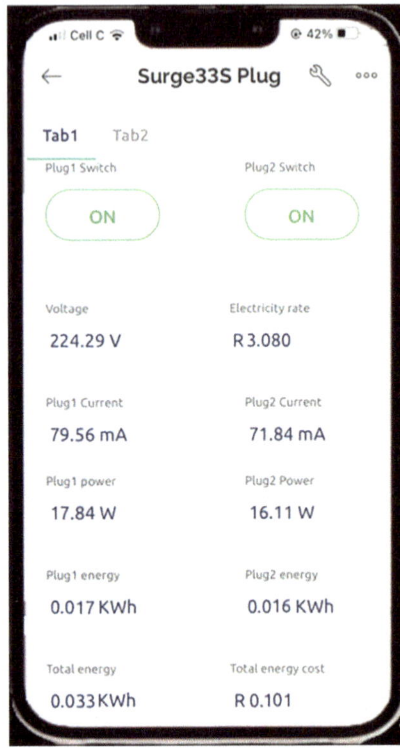

Fig. 8 Smartphone face

3.4 *Hardware Design*

Transitioning to the tangible aspects of the hardware design of the smart plug prototype, the prototype development goes on to deal with the physical construction and assembly that brings the project to life. The process began by designing the enclosure of the smart plug to house the components using Autodesk Fusion 360 (Fig. 10).

Figure 10 illustrates the measurements that were taken for the proper fitting of the components. The sketch was then printed out using a 3D machine. Following the successful 3D printing of the enclosure, the process carried by meticulously arranging the components, conducting vital drilling, skillful soldering, and precise wiring to bring the hardware of the smart plug prototype to its completion (Fig. 11).

Briefly, the hardware design represents the solid foundation of our smart plug system, where each component's placement and integration have been carefully orchestrated. This well-engineered physical framework ensures efficient operation, user friendliness, and, above all, paves the way for seamless energy monitoring and control.

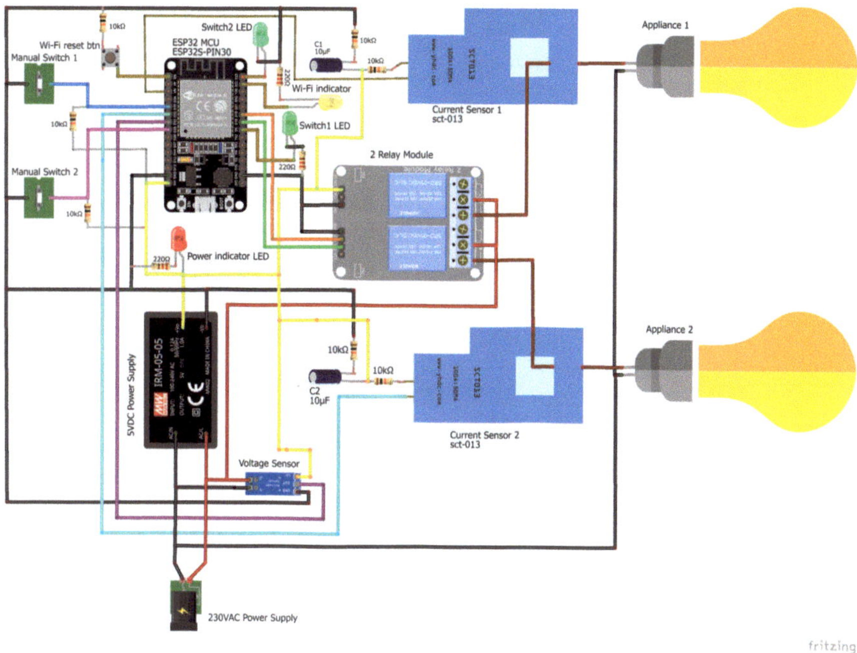

Fig. 9 Schematic circuit

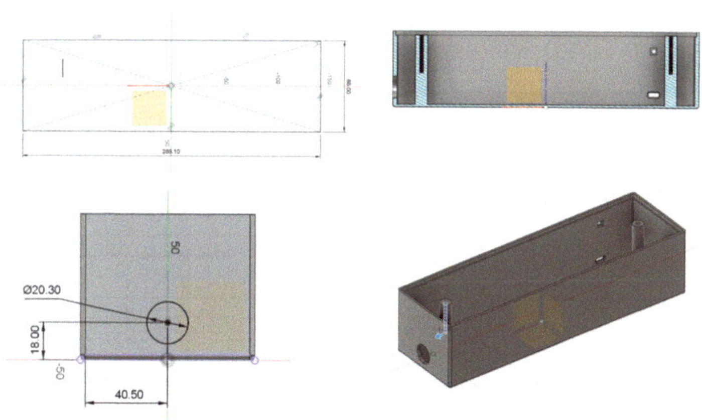

Fig. 10 3D design of the smart plug enclosure

3.5 *Software and Control App Design*

This section deals with the software design aspect of the project and uncovers how the software interfaces with the hardware by adding an essential layer of functionality. The block diagram presented in Fig. 12 summarizes the comprehensive algorithm that enables the smart plug to operate at its full potential.

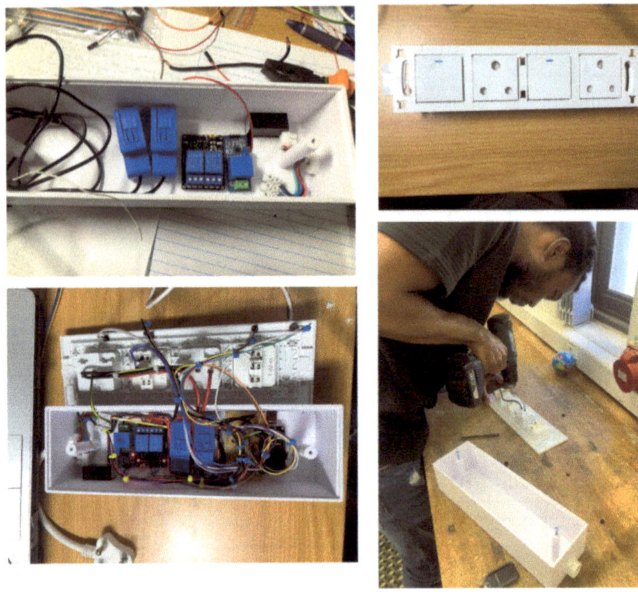

Fig. 11 Hardware development

The system begins by configuring various parameters and establishing a connection with the Blynk app, enabling remote control. The voltage sensor (ZMPT01B) and current sensor (SCT013) are initialized for accurate energy monitoring. In the main loop, the system continuously checks the state of physical switches to ensure synchronization with relay states and updates the Blynk app accordingly. Additionally, a timer triggers plug calculations every 10 seconds, acquiring data on voltage, current, power, and energy usage. This information is displayed on both the Blynk app dashboard, and the serial monitor for user reference and system debugging. The system ensures efficient energy management and user-friendly control, both locally and remotely. The detailed algorithm processed by the ESP32 is shown in Fig. 13.

3.6 ESP32 Software Development

The smart plug algorithm begins by powering the ESP32 microcontroller, which serves as the core component enabling both control and data processing. Once powered, the ESP32 checks for a Wi-Fi connection. If connected, the device transitions to online mode, intergrating seamlessly with the Blynk mobile application. In this mode, the smart plug connects an appliance, such as a household device, whose operating voltage and current draw are measured by the ESP32 via its ADC pins. These readings are used to calculate real-time power and energy consumption,

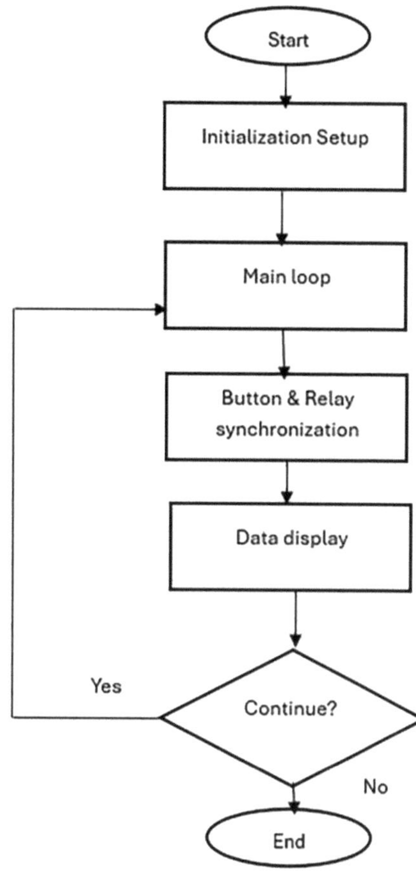

Fig. 12 Software algorithms overview

which are then transmitted to the Blynk server using WebSocket protocol. WebSocket is ideal for this application as it establishes a persistent, full-duplex communication channel between the ESP32 and the Blynk cloud, ensuring low-latency updates of real-time data such sensor readings and control commands. Users can remotely monitor and control the smart plug, toggling the connected appliance ON or OFF directly from the Blynk app.

If the ESP32 does not detect a Wi-Fi connection, it enters Access Point (AP) mode, creating a local Wi-Fi network for configuration. Users can connect to this network, input Wi-Fi credentials via a browser interface, and save these details to the ESP32's EEPROM memory. The device then restarts and attempts to reconnect. In cases where Wi-Fi is unavailable, the smart plug switches to offline mode, providing basic functionality such as manual control. This system bridges household appliances to the Internet of Things by utilizing Wi-Fi for connectivity and WebSocket for real-time data exchange, ensuring seamless communication, energy monitoring, and remote control via the Blynk app.

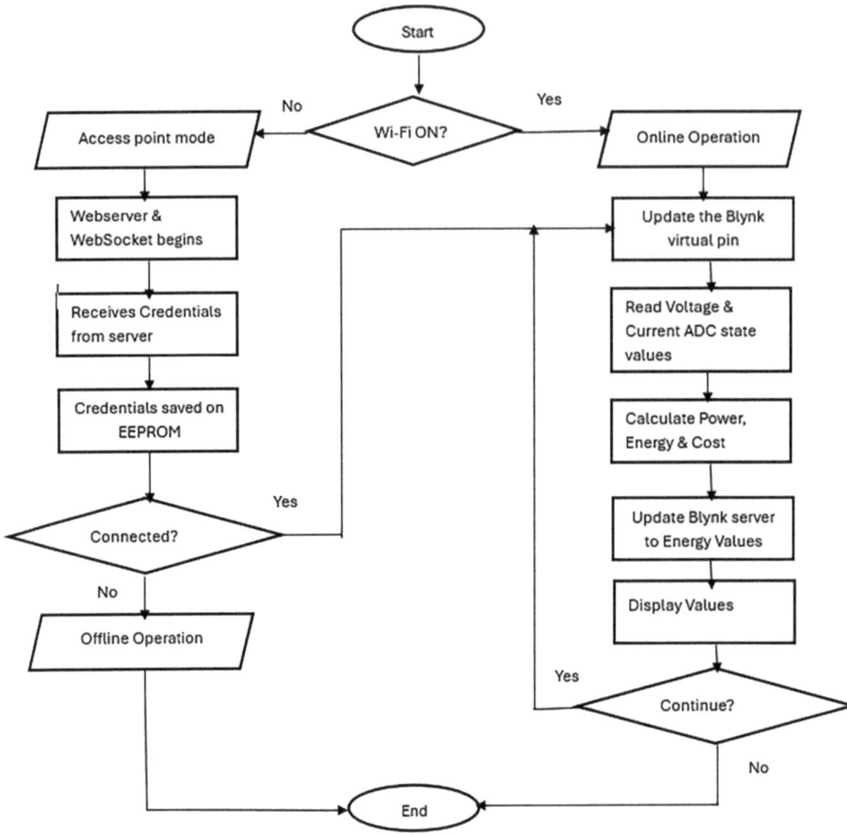

Fig. 13 System detailed algorithm

3.7 Testing and Calibration of Sensors

The testing and calibration of sensors in the smart plug prototype were critical steps in ensuring accurate and reliable data. The voltage and current sensors were rigorously tested by comparing their measurements with known reference values to verify precision and suitability for the application. Calibration was an essential process during which sensors were carefully adjusted to ensure consistent readings that matched real-world measurements. Calibration factors were meticulously fine-tuned to account for any discrepancies.

Figure 14 illustrates a sine wave on the right and the corresponding voltage reading of 236.3V on the left, displayed on the TRMS multimeter, indicating the successful calibration of the voltage sensor. The calibration of the current sensor was performed within the code.

Throughout the testing and calibration phases, a record of data and adjustments were kept, ensuring the highest level of accuracy in measurements. These processes

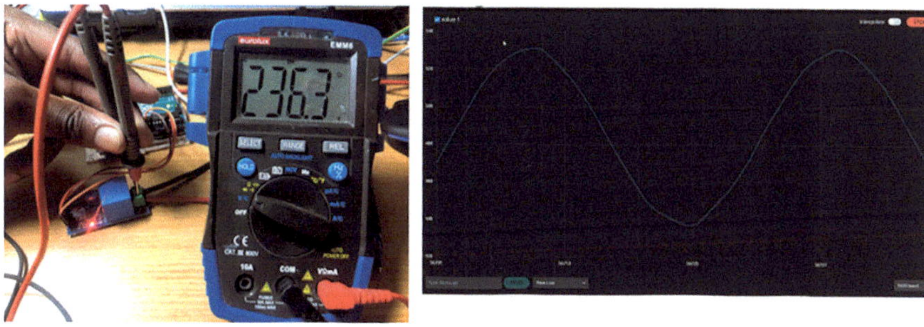

Fig. 14 Capture for the voltage sensor calibration

were fundamental in guaranteeing that the data collected and presented to users is reliable and trustworthy, making the smart plug an indispensable tool for energy monitoring and control.

3.8 Operation of the Smart Plug Prototype

This section discusses the practical operation of the prototype, providing insights into how users interact with the system and how it seamlessly fulfills its intended functions. The smart plug, powered by a robust digital infrastructure, presents a user-friendly feature. When plugged into an electrical socket, users can manually control the connected appliances through physical switches synchronized with a corresponding mobile application. With just a few taps on their smartphones, users can remotely turn appliances on or off, enabling real-time control. The smart plug provides essential energy data, displaying metrics such as voltage, current, power, energy consumption, and cost on the Blynk app. This information empowers users to monitor and optimize their energy usage, promoting efficiency and cost savings. The inclusion of energy cost calculations further enhances the plug's practicality. Through straightforward operation, the smart plug simplifies energy management by providing a convenient and effective solution for households and businesses alike.

3.9 System Integration and Prototype Results

After completing the design and development of the smart plug prototype, a series of tests and assessments evaluated its functionality and effectiveness. The results proved both promising and significant, as discussed in the next section. Figure 15 depicts a working smart networked plug with its corresponding data shown on the Blynk app on the cellphone.

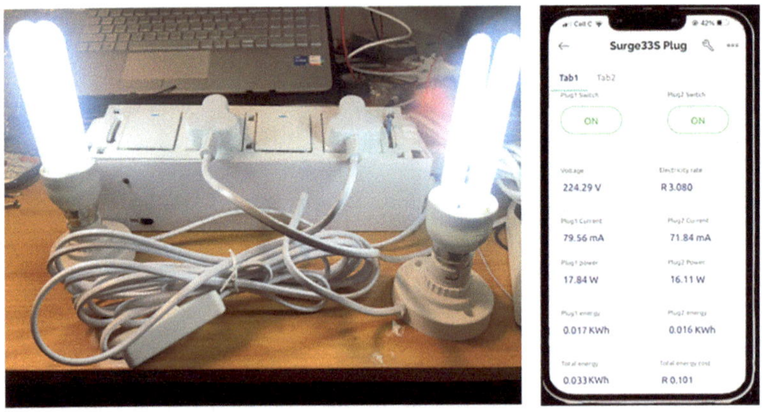

Fig. 15 Networked plug smart metering reading display

4 Smart Meter Performance Evaluation

The performance of the smart meter was established comparing the data collected empirically against that of the customer-end smart meter.

4.1 Data Collection

For the empirical data collection, the data was collected from six appliances using a T235H digital multimeter (DMM) to measure the supply voltage and a TBM303 clamp-on digital multimeter to measure the supply current. The DMM was chosen because the academic project relied on school lab equipment, and this device was the available tool. Power was calculated using Eq. (1). Appliances were connected for a duration of one hour, and energy was calculated according to Eq. (2).

$$P = VI \tag{1}$$

where P is the power, V is the voltage, and I is the current

$$E = Pt \tag{2}$$

where E is the energy and t is the time the appliance has spent connected to the networked plug. The energy cost was calculated based on the energy tariff of the day in South Africa which was R3.08/kWh.

For the proposed smartphone-based meter, the networked smart plug wirelessly transmits voltage and current measurements to the cellphone via the Blynk

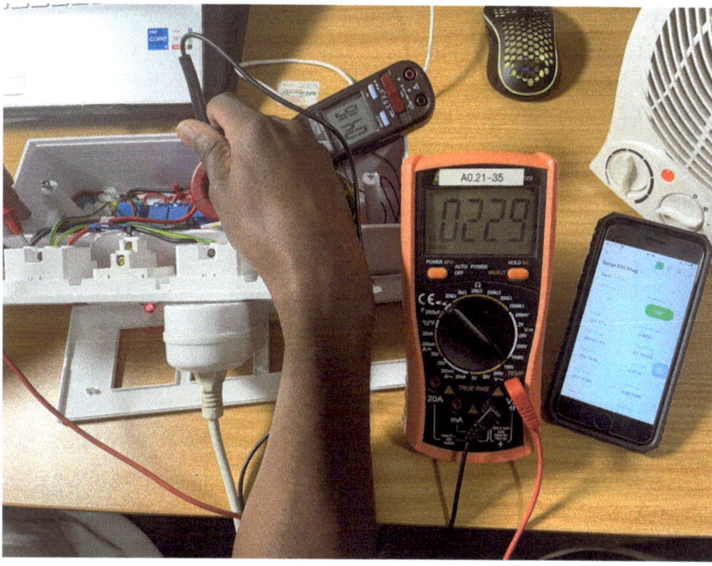

Fig. 16 Measurement set up for a fan heater

Table 1 Manually measured data and calculated on five appliance over 1 h

Appliances	Voltage (V)	Current (A)	Power (kW)	Time (h)	Total energy (kWh)	Cost per unit (R)	Total cost (R)
1. Fan heater	233	4.24	0.988	1	0.988	3.08	3.04
2. Iron	236	6.61	1.56	1	1.56	3.08	4.8
3. Light bulb	236	0.08	0.019	1	0.019	3.08	0.06
4. Hot plate	230	4.57	1.051	1	1.051	3.08	3.24
5. Fridge	236	0.49	0.116	1	0.116	3.08	0.36
6. Laptop	236	0.2	0.047	1	0.047	3.08	0.14

application. The algorithm within the Blynk app simultaneously computes the power, energy consumption, and the cost, and displays the results on the smartphone screen. Figure 16 presents data collected from a fan heater. A digital multimeter, a current clamp-on meter, and smartphone running the Blynk app collect the measurements. The results for empirically measured and calculated data are displayed in Table 1, while those for the smart plug are shown in Table 2. Tables 3 and 4 provide computations of means and variance parameters for DMM-measured and smart metering measured systems, respectively.

Table 2 Data collected with a smartphone running the Blynk application

Appliances	Voltage (V)	Current (A)	Power (kW)	Time (h)	Total energy (kWh)	Cost per unit (R)	Total cost (R)
1. Fan heater	231.2	3.9	0.902	1	0.902	3.08	2.78
2. Iron	234.12	6.45	1.51	1	1.51	3.08	4.65
3. Light bulb	230.4	0.079	0.018	1	0.018	3.08	0.06
4. Hot plate	229.6	4.21	0.967	1	0.967	3.08	2.98
5. Fridge	235.7	0.42	0.099	1	0.099	3.08	0.3
6. Laptop	233.18	0.17	0.04	1	0.04	3.08	0.12

Table 3 Average values and variance for manually measured quantities

	Voltage (V)	Current (A)	Power (kW)	Energy (kWh)	Cost (R)
Average	234.5	2.698	0.63	0.63	1.94
Variance	6.3	7.831	0.430	0.430	4.070

Table 4 Average values and variance for smart-metering measured quantities

	Voltage (V)	Current (A)	Power (kW)	Energy (kWh)	Cost (R)
Average	232.367	2.53	0.589	0.589	1.815
Variance	5.546	7.219	0.391	0.391	3.71479

4.2 Results and System Performance Analysis

With the values of quantities of interest obtained, it is necessary to establish the similarity between the two data sets- empirically measured and smart system-collected. Comparing the mean values is the most efficient approach. The empirically measured and calculated data will serve as the reference for assessing the performance of the proposed system.

4.2.1 Variance Comparison

The study of mean similarity can be explored through the equality of two dataset variances, determined by calculating the variance ratio [27]. Table 5 presents the variances of the two datasets and the ratio between them.

The computation of variance ratios (smart over manual) for different parameters yields values of aproximately 0.9 which is close to 1 for all the parameters-indicating that they are almost equal. However, the means still need to be compared using hypothesis testing for the mean values of voltages, currents, energies, and the costs between empirically measured and smartly obtained data.

Table 5 Comparison of variances between manual and smart metering

Metering system	Voltage (V)	Current (A)	Energy (kWh)	Cost (R)
DMM	6.3	7.831	0.430	4.070
Smart	5.546	7.219	0.391	3.715
Ratio	0.880 (88%)	0.922 (92%)	0.911 (91%)	0.913 (91%)

4.2.2 Hypothesis Testing

In this case, the null hypothesis is: "*The mean value for a manually measured parameter equals the mean value of the smartly measured parameter*." The alternate hypothesis is: "*The mean value for a parameter is different for the two metering systems*." Mathematically, this can be represented as follows (Eqs. 3 and 4).

$$H_0 : \mu_1 = \mu_2 \tag{3}$$

$$H_1 : \mu_1 \neq \mu_2 \tag{4}$$

In the process to conduct the hypothesis testing, we resort to statistics inferences by establishing the acceptability and the rejectability of the null hypothesis [28]. For this, we will use a confidence interval of 95% and a two-tail t-distribution. Since the size of each of the two data sets is 6, the degrees of freedom is expressed as (Eq. 5):

$$\mathrm{DF} = n_1 - 1 + n_2 - 1 \tag{5}$$

This implies that the degrees of freedom is 10. Scanning the student's t-distribution for critical values with the 95% confidence interval of the two-tail distribution, $t_{\mathrm{Critical}} = 2.228$.

Therefore, any calculated t value that is greater than this will lead to the rejection of the null hypothesis, hence to the disqualification of the proposed smart metering system (Fig. 17).

4.2.3 *t*-Value Computing

The t-value is computing using this formula for two data sets (Eq. 6):

$$t = \frac{\bar{x}_1 - \bar{x}_2}{S_\mathrm{p}\sqrt{\frac{1}{n_1} + \frac{1}{n_2}}} \tag{6}$$

where n_1 and n_2 are the number of samples and (Eq. 7)

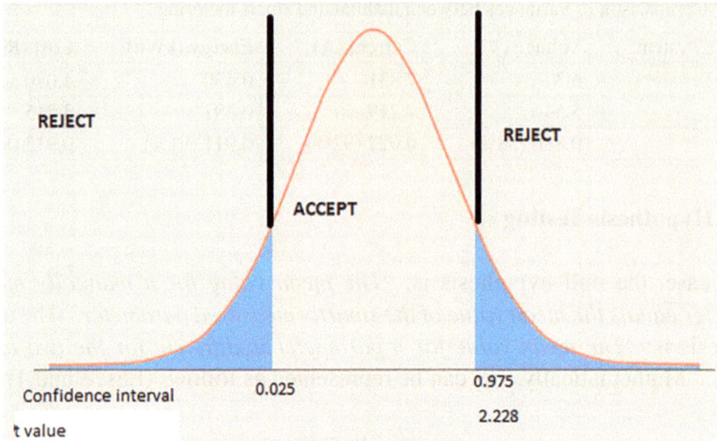

Fig. 17 Hypothesis acceptance and rejection zone for a 10 degrees of freedom

Table 6 *t*-Score values

	Voltage	Current	Power	Cost
t-Score	−1.518	−0.248	−0.272	−0.269

$$S_p = \sqrt{\frac{(n_1 - 1)s_1^2 + (n_2 - 1)s_2^2}{(n_1 - 1 + n_2 - 2)}} \quad (7)$$

where $\overline{x}_1$ and $\overline{x}_2$ are the means of the first and second data sets, respectively, s_1 and s_2 are their standard deviations, S_p is the pooled standard deviation for the two data sets.

Using Eq. (5) and the results in Table 5, we obtain *t* scores as per Table 6.

The *t* values computed are negative because mean values from the smart plug are lower than those measured by the DMM. These values should be treated as positive and compared to the critical *t* value. All computed values are lower than the critical value of 2.228, indicating that the null hypothesis -that the mean values are equal-holds true for voltage, current, power, and cost. Therefore, the performance of the proposed end-user smart metering system is commendable.

4.3 *Study for the Probability Density Function*

The study of the probability density function (PDF) of the data set provides the occurrence probability for random data. Hence, the data set with a high PDF is understood to have the highest likelihood occurrence. The expression of the PDF is per Eq. (8).

$$\mathrm{PDF} = \int_a^b f(x)\,dx \tag{8}$$

where $f(x)$ is expressed by Eq. (9)

$$f(x) = \frac{1}{\sigma\sqrt{2\pi}} e^{-\frac{1}{2}\left(\frac{x-\mu}{\sigma}\right)^2} \tag{9}$$

where x is the random data in the data set, μ is the mean, and σ is the standard deviation for the data set.

In the case of this study, MATLAB was used to compute the PDF and for various data sets based on the lower and upper limits and the mean and standard deviation of the data sets. PDF graphs were also captured as per Fig. 18.

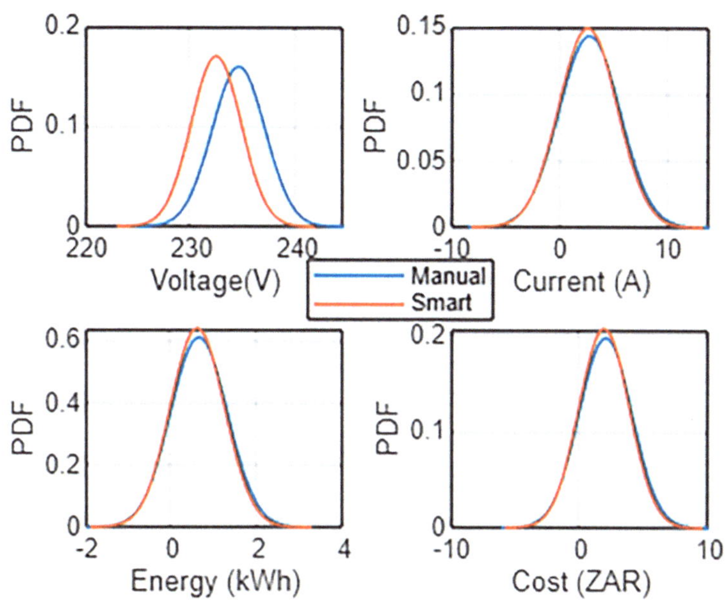

Fig. 18 Probability density function for all parameters for the two approaches

The code written to generate the PDF plots is as follows:

```
%**********************************************
% Code to Calculate the Probability Density Function
%Author: Janvier Kamanzi
%Date: December 2024
%**********************************************
clear
close all
clc

average1 = [234.5 2.698 0.63 1.94];
std1 = sqrt([6.3 7.83 0.43 4.1]);

average2 = [232.367 2.53 0.589 1.82];
std2 = sqrt([5.546 7.219 0.391 3.714]);
limitL1 = average1 - 4 * std1;
limitU1 = average1 + 4 * std1;

limitL2 = average2 - 4 * std2;
limitU2 = average2 + 4 * std2;

v1= linspace(limitL1(1), limitU1(1));
vp1 = normpdf(v1, average1(1), std1(1));

c1= linspace(limitL1(2), limitU1(2));
cp1 = normpdf(c1, average1(2), std1(2));

v2= linspace(limitL2(1), limitU2(1));
vp2 = normpdf(v2, average2(1), std2(1));

c2= linspace(limitL2(2), limitU2(2));
cp2 = normpdf(c2, average2(2), std2(2));

e1= linspace(limitL1(3), limitU1(3));
ep1 = normpdf(e1, average1(3), std1(3));

e2= linspace(limitL2(3), limitU1(3));
ep2 = normpdf(e2, average2(3), std2(3));

cost1= linspace(limitL1(4), limitU1(4));
costp1 = normpdf(cost1, average1(4), std1(4));

cost2= linspace(limitL2(4), limitU2(4));
costp2 = normpdf(cost2, average2(4), std2(4));
```

```
subplot(2,2,1)
    plot(v1, vp1, v2, vp2)
    legend('Manual', 'Smart')
    xlabel('Voltage(V)')
    ylabel('PDF')
    grid on
subplot(2,2,2)
    plot(c1, cp1, c2, cp2)
    xlabel('Current (A)')
    ylabel('PDF')
    grid on
subplot(2,2,3)
    plot(e1, ep1, e2, ep2)
    xlabel('Energy (kWh)')
    ylabel('PDF')
    grid on
    subplot(2,2,4)
    plot(cost1, costp1, cost2, costp2)
    grid on
    ylabel('PDF')
    xlabel('Cost (ZAR)')
%***********************************************
```

The reflection of PDF results in Fig. 18 leads to the following conclusion: the data set displayed by the smart plug exhibits higher PDFs than the data collected manually using the DMMs and is closer to the mean. Furthermore, the proposed smart plug readings provide data with greater precision, particularly for larger quantities such as voltage.

5 Conclusion

The proposed end-user smart metering system effectively provides real-time energy monitoring and remote switching of appliances. Users can seamlessly control appliances from their smartphones and access critical data on voltage, current, power consumption, energy usage, and cost. This capability enables informed decision-making, helps identify energy-intensive appliances, and supports measures to reduce electricity bills.

The prototype's power calculations are highly accurate, due to precise calibration of voltage and current sensors, ensuring reliable data for effective energy management. The inclusion of energy cost calculations is a notable feature major highlight. It enables users to understand the financial implications of their energy usage, promoting conscious consumption and cost-saving pr, allowing users to assess the financial impact of their energy usage, thereby promoting consious consumption and cost-saving practices. Additionally, the mobile application user-friendly interface simplifies operation, making energy management accessible to a wide range of users. The ability to monitor two different appliances simultaneously, with clearly presented data for each on the Blynk app, enhances practicality.

While this work focuses on home applications, future developments aim to expand its scope. Planned enhancements include leveraging big data, Artificial Intelligence (AI) and Machine Learning (ML) to further improve functionality and adaptability.

References

1. Yalli, J.S., Hasan, M.H., Badawi, A.: Internet of Things (IOT): origin, embedded technologies, smart applications and its growth in the last decade. IEEE Access, 91357–91382 (2024). https://doi.org/10.1109/ACCESS.2024.3418995
2. Dhaou, I.B.: Design and implementation of an Internet-of-Things-enabled smart meter and smart plug for home-energy-management system. Electronics (Switzerland). **12**, 1–20 (2023). https://doi.org/10.3390/electronics12194041
3. Santos, A., Duggan, G.P., Davis, J., Zimmerle, D.: A cautionary note on using smart plugs for research data acquisition. e-Prime Adv. Electr. Eng. Electron. Energy. **4** (2023). https://doi.org/10.1016/j.prime.2023.100137
4. Laktionov, I., Diachenko, G., Kashtan, V., Vizniuk, A., Gorev, V., Khabarlak, K., Shedlovska, Y.: A comprehensive review of recent approaches and hardware-software technologies for digitalisation and intellectualisation of open-field crop production: Ukrainian case study in the global context. Comput. Electron. Agric. **225**, 109326 (2024). https://doi.org/10.1016/j.compag.2024.109326
5. Kurni, M., Srinivasa, K.G.: IoT-based educational software and hardware systems. In: The Internet of Educational Things: Enhancing Students' Engagement and Learning Performance (2025)
6. Sanchez-Sutil, F., Cano-Ortega, A.: Smart plug for monitoring and controlling electrical devices with a wireless communication system integrated in a LoRaWAN. Expert Syst. Appl. **213**, 118976 (2023). https://doi.org/10.1016/j.eswa.2022.118976
7. Singh, S., Chavekar, A., Mane, S., Kazi, F.: Exploring chip-off firmware extraction techniques and challenges: case studies in smart plugs. In: Proceedings of the International Carnahan Conference on Security Technology, pp. 1–6 (2023). https://doi.org/10.1109/ICCST59048.2023.10726849
8. Jafarov, N., Nasiyev, M.: Analysis of the Lora Wan structure and advantages of it over traditional IoT networks, pp. 59–64 (2023)
9. Lu, Y., Da Xu, L.: Internet of Things (IoT) cybersecurity research: a review of current research topics. IEEE Internet Things J. **6**, 2103–2115 (2019). https://doi.org/10.1109/JIOT.2018.2869847
10. Lee, I.: Internet of Things (IoT) cybersecurity: literature review and IoT cyber risk management. Future Internet. **12** (2020). https://doi.org/10.3390/FI12090157
11. Lee, S.H., Yang, C.S.: An intelligent power monitoring and analysis system for distributed smart plugs sensor networks. Int. J. Distrib. Sens. Netw. **13**, 155014771771846 (2017). https://doi.org/10.1177/1550147717718462
12. Ling, Z., Luo, J., Xu, Y., Gao, C., Wu, K., Fu, X.: Security vulnerabilities of Internet of Things: a case study of the smart plug system. IEEE Internet Things J. **4**, 1899–1909 (2017). https://doi.org/10.1109/JIOT.2017.2707465

13. Stan, C.: Current energy scenarios: decarbonization. In: Energy Scenarios for the Future: Ways to Climate-friendly Mobility, Heating and Industry. Springer (2024)
14. Alhasnawi, B.N., Almutoki, S.M.M., Hussain, F.F.K., Harrison, A., Bazooyar, B., Zanker, M., Bureš, V.: A new methodology for reducing carbon emissions using multi-renewable energy systems and artificial intelligence. Sustain. Cities Soc. **114**, 105721 (2024). https://doi.org/10.1016/j.scs.2024.105721
15. Elia, A., Kamidelivand, M., Rogan, F., Gallachóir, B.Ó.: Impacts of innovation on renewable energy technology cost reductions. Renew. Sust. Energ. Rev. **138**, 110488 (2021). https://doi.org/10.1016/j.rser.2020.110488
16. Goudarzi, A., Ghayoor, F., Waseem, M., Fahad, S., Traore, I.: A survey on IoT-enabled smart grids: emerging, applications, challenges, and outlook. Energies. **15**, 6984 (2022)
17. Förderer, K., Grafenhorst, S., Hagenmeyer, V.: Distribution Grid Monitoring Based on Widely Available Smart Plugs System Platform for Borehole Tools View Project SmartMES View Project. https://doi.org/10.48550/arXiv.2211.04211
18. Salih, R.: Adagio for the Internet of Things. IoT Penetration Testing and Security Analysis of a Smart Plug (2021)
19. Yu, X., Dong, Z., Zheng, D., Deng, S.: Analysis of critical peak electricity price optimization model considering coal consumption rate of power generation side. Environ. Sci. Pollut. Res. **31**, 41514–41528 (2024). https://doi.org/10.1007/s11356-023-29754-5
20. Ayan, O., Turkay, B.E.: Smart home energy management technologies based on demand side management and a review of smart plugs with smart thermostats (2017)
21. Erdin, C., Ozkaya, G.: Turkey's 2023 energy strategies and investment opportunities for renewable energy sources: site selection based on ELECTRE. Sustainability (Switzerland). **11** (2019). https://doi.org/10.3390/su11072136
22. Cho, Y., Choi, A.: A method to facilitate affordance perception and actualization for improving the usability of smart plugs. Build. Environ. **217**, 109095 (2022). https://doi.org/10.1016/j.buildenv.2022.109095
23. Baudot, L., Robson, K.: Regulation. The Routledge Companion to Critical Accounting, vol. 2014, pp. 184–204. Routledge (2017). https://doi.org/10.4324/9781315775203-11
24. Preview, T.S.: International Standard Norme Internationale Electricity Metering Equipment-Particular Requirements-Part 21: Static Meters for AC Active Energy (Classes 0.5, 1 and 2) (2020)
25. Admin: Power ESP8266 with HLK-PM03 Converter | Random Nerd Tutorials. https://randomnerdtutorials.com/esp8266-hi-link-hlk-pm03/
26. Admin: IoT Based Electricity Energy Meter using ESP32 & Blynk. https://how2electronics.com/iot-based-electricity-energy-meter-using-esp32-blynk/
27. Zhang, J.T.: Analysis of Variance for Functional Data. Routledge (2013)
28. Klein, J.P., Moeschberger, M.L.: Hypothesis testing. In: Survival Analysis: Techniques for Censored and Truncated Data. Springer (2003)

Recent Developments in Artificial Intelligence (AI) Sensors

Weston Mwashita

1 Introduction

Artificial Intelligence (AI) has rapidly transformed various industries, with sensors playing a crucial role in enabling intelligent systems to perceive, process, and respond to their environment. The integration of AI with sensor technology has led to the development of advanced systems that can collect, analyse, and act upon data in real time. AI sensors have seen remarkable improvements in areas such as accuracy, response time, and decision-making capabilities, driving innovation in fields like healthcare, manufacturing, robotics, and autonomous systems [1]. These developments are particularly important as AI-driven systems rely on high-quality data inputs from sensors to ensure efficient and accurate outcomes.

One of the most significant advancements in AI sensors is their ability to perform edge computing, where data processing occurs at the sensor itself, reducing latency and improving response times [2]. Edge AI sensors are particularly beneficial in applications where real-time decision-making is crucial, such as autonomous vehicles and industrial automation [3]. This shift toward decentralized processing has also alleviated the burden on cloud computing infrastructure, allowing systems to operate independently and efficiently without the need for constant connectivity [4].

In healthcare, AI sensors are revolutionizing patient monitoring and diagnosis. Wearable sensors integrated with AI algorithms can now continuously monitor vital signs, detect abnormalities, and predict health issues in real time [5]. Such sensors not only improve patient outcomes but also reduce the workload on medical professionals, allowing for more accurate and timely interventions. Similarly, in smart manufacturing, AI-enhanced sensors are enabling predictive maintenance by

W. Mwashita (✉)
International University of Management, Windhoek, Namibia
e-mail: w.mwashita@ium.edu.na

P. Naidoo, M. Sibanda (eds.), *Measurement, Automation, and Control in Artificial Intelligence and Machine Learning*, Signals and Communication Technology, https://doi.org/10.1007/978-3-031-91802-5_10

analysing machinery performance data and predicting potential failures before they occur, thus minimizing downtime and optimizing operational efficiency [6].

Despite these advancements, challenges remain, particularly in the areas of sensor data accuracy, energy efficiency, and security. As AI sensors continue to evolve, ongoing research and development are focused on overcoming these limitations while expanding their applications in a variety of industries. This chapter explores the latest innovations in AI sensor technology, highlights the breakthroughs achieved, and discusses the future directions of this rapidly growing field.

2 AI Sensors: A Technological Evolution

The incorporation of AI into sensors has shifted their role from simple data-gathering devices to smart systems capable of autonomous decision-making. One key advancement is the ability of AI sensors to perform complex tasks such as pattern recognition, anomaly detection, and predictive analysis without human intervention. Traditional sensors, which provided raw data, now benefit from embedded AI algorithms that enable them to interpret the information collected. This ability to 'sense' and 'think' has opened doors to various applications [7].

AI sensors now play a critical role in autonomous vehicles, where they manage multiple streams of data from cameras, radars, and LIDAR systems. Through AI, these sensors not only detect objects but also predict their behaviour, allowing vehicles to navigate safely and efficiently in real time [8]. The integration of AI has similarly revolutionized industrial automation by enabling sensors to monitor machine conditions and predict equipment failures before they occur, which has drastically reduced downtime and maintenance costs [9].

2.1 Breakthroughs in Sensor Data Accuracy

One of the major challenges in AI sensors is data accuracy, particularly in noisy or dynamic environments. AI algorithms are vulnerable to erroneous outputs if the input data is corrupted or incomplete. Recent advancements have focused on improving sensor data accuracy through noise reduction techniques, adaptive learning algorithms, and fusion of data from multiple sensor modalities [10].

A notable innovation in this area is the use of deep learning to enhance data quality in image and audio sensors. Convolutional neural networks (CNNs) have been deployed to clean noisy data, identify patterns, and provide more accurate recognition capabilities. In autonomous driving, for instance, AI sensors using multi-sensor fusion can combine inputs from cameras, radar, and LIDAR to achieve higher accuracy in detecting obstacles and determining distances, thereby improving decision-making [11].

Further, advanced error-correction algorithms have been developed to mitigate inaccuracies in environmental and industrial applications, where sensors are exposed to harsh conditions. These algorithms enable sensors to self-calibrate and adapt to fluctuating conditions, ensuring that the data remains reliable [12].

2.2 *Energy Efficiency in AI Sensors*

Energy efficiency in Artificial Intelligence (AI) sensors has become a critical area of focus as these sensors become more pervasive across industries such as healthcare, manufacturing, smart cities, and autonomous systems. AI sensors are responsible for data collection, processing, and communication, all of which require energy. Given the rising importance of sustainability and the increasing deployment of sensors in power-constrained environments like Internet of Things (IoT) networks, enhancing the energy efficiency of AI sensors is crucial. This section explores various approaches to improving energy efficiency in AI sensors, highlighting both their strengths and weaknesses, with real-world examples where applicable.

2.2.1 Low-Power AI Algorithms

One of the key approaches to improving energy efficiency in AI sensors is the development of low-power AI algorithms. These algorithms are designed to reduce the computational load on the sensor, thereby minimizing energy consumption. AI algorithms typically consume substantial energy, especially when applied to large datasets or complex models. Low-power algorithms optimize performance by simplifying operations, reducing the number of computations, or using specialized hardware such as low-power microcontrollers.

In the context of wearable health devices, where continuous data collection is essential, low-power AI algorithms play a pivotal role. Wearable AI sensors, such as those monitoring heart rates, blood oxygen levels, or physical activity, rely on these algorithms to extend battery life without compromising accuracy. For example, research has shown that the implementation of low-power AI algorithms in wearable health sensors reduced power consumption by up to 40%, enabling longer operational times between charges [13].

Strengths

- Low-power algorithms significantly reduce energy consumption, which extends the battery life of devices and reduces the need for frequent recharging.
- They can be deployed in various environments, from personal devices to industrial settings, where energy efficiency is paramount.
- These algorithms can adapt to different operating conditions, making them flexible and versatile.

Weaknesses

- Simplified algorithms may compromise accuracy and performance, particularly in applications that require high precision.
- Developing low-power algorithms requires specialized knowledge of both hardware and AI models, making them more challenging to implement.
- There may be limitations in their scalability when applied to complex, data-intensive AI tasks [14].

2.2.2 Energy Harvesting for AI Sensors

Energy harvesting technologies offer a promising solution to the energy efficiency challenge by enabling AI sensors to generate their own power from ambient sources such as solar energy, thermal gradients, vibration, or radio frequency (RF) waves. These technologies allow AI sensors to operate independently of external power supplies, reducing the need for batteries or frequent recharging.

An example of energy harvesting in action is in smart agriculture, where AI sensors monitor environmental conditions such as soil moisture, temperature, and sunlight exposure. Solar-powered AI sensors are particularly useful in remote or difficult-to-access areas where battery replacement is impractical. By harvesting solar energy, these sensors can operate for extended periods without human intervention. Recent advancements in energy harvesting techniques have enabled sensors to become more energy-efficient, with some models capable of harvesting energy even under low-light conditions [15].

Strengths

- Energy harvesting reduces dependency on batteries, which can lower maintenance costs and improve the sustainability of sensor networks.
- It enables sensors to operate in remote or difficult-to-reach locations without requiring regular maintenance or power supplies.
- Energy harvesting technologies are environmentally friendly, reducing the carbon footprint of large-scale sensor networks.

Weaknesses

- The amount of energy harvested is often limited, making it unsuitable for sensors with high power demands or in environments with low energy availability (e.g., low-light or low-vibration areas).
- Energy harvesting systems are often more expensive and complex to design, particularly when integrating multiple energy sources (solar, thermal, etc.).
- Environmental conditions, such as weather or the availability of ambient energy sources, can affect the reliability of these systems [16].

2.2.3 Edge Computing and Data Compression Techniques

Edge computing has emerged as an effective way to enhance energy efficiency in AI sensors by reducing the need for data transmission to central servers or cloud platforms. In traditional IoT setups, sensors collect data and transmit it to a central system for processing, which requires significant energy for both communication and computation. Edge computing reduces this energy consumption by processing data locally at the sensor or near the sensor node, thereby minimizing communication overhead and reducing latency.

One notable application of edge computing is in autonomous vehicles. These vehicles rely on multiple AI sensors, such as cameras, LiDAR, and radar, to gather data in real time. Instead of transmitting all the data to a remote server, edge computing allows data processing to be done locally within the vehicle, reducing energy consumption while improving responsiveness. Studies have shown that edge computing can reduce the energy consumption of AI sensors in autonomous systems by up to 30% [17].

In addition to edge computing, data compression techniques are often used to reduce the size of the data transmitted, thereby saving energy. Data compression algorithms analyse the raw sensor data and compress it into smaller, more manageable packets, minimizing the energy required for transmission while preserving the essential information.

Strengths

- Edge computing significantly reduces communication energy costs, which can be substantial in large sensor networks or systems with limited bandwidth.
- It enables real-time decision-making by processing data locally, reducing latency, and improving system responsiveness.
- Data compression can be applied to a wide range of sensors, reducing the energy required for data transmission without major modifications to the sensor hardware.

Weaknesses

- Implementing edge computing requires additional processing power at the sensor level, which can increase the cost and complexity of the sensor node.
- Data compression may result in some loss of information, which can affect the accuracy of AI models, particularly in applications where precise data is critical.
- Edge computing may not be suitable for all applications, especially in systems that require centralized data processing for complex tasks [18].

2.2.4 Low-Power Hardware for AI Sensors

Another important factor in improving the energy efficiency of AI sensors is the development of low-power hardware. This includes the design of energy-efficient microcontrollers, application-specific integrated circuits (ASICs), and field-programmable gate arrays (FPGAs) that are optimized for AI computations. These hardware components are designed to execute AI algorithms more efficiently, reducing the overall power consumption of the sensor system.

For instance, AI-powered image sensors used in surveillance or industrial monitoring can benefit from low-power hardware solutions. Recent developments in low-power microcontrollers have enabled AI sensors to process large amounts of image data while consuming only a fraction of the energy compared to traditional processors. One study demonstrated that the use of low-power microcontrollers reduced the energy consumption of AI image sensors by up to 50%, making them more suitable for long-term deployments in energy-constrained environments [19].

Strengths

- Low-power hardware significantly reduces the energy consumption of AI sensors, allowing for longer operational lifespans and more efficient systems.
- These hardware solutions are optimized for specific AI tasks, which enhances both energy efficiency and performance.
- Low-power hardware can be integrated into a wide range of sensor systems, from wearable devices to industrial automation systems.

Weaknesses

- Low-power hardware can be expensive to develop and implement, particularly in custom AI sensor applications.
- These solutions may require specialized design and manufacturing processes, which can limit their scalability in certain industries.
- Hardware upgrades may be needed as AI models become more complex, increasing the cost and effort associated with maintaining energy-efficient systems [7].

2.2.5 Challenges and Future Directions

While significant progress has been made in improving the energy efficiency of AI sensors, several challenges remain. One of the primary issues is the trade-off between energy efficiency and performance. Many of the techniques described above, such as low-power algorithms and data compression, can result in reduced accuracy or slower processing times, which may not be acceptable in certain critical applications. Additionally, the integration of energy harvesting technologies and low-power hardware requires careful consideration of cost, complexity, and scalability.

Future research in this area is focused on developing hybrid solutions that combine multiple energy-saving techniques, such as energy harvesting and low-power hardware, while optimizing AI algorithms for minimal energy use. There is also growing interest in exploring neuromorphic computing, which mimics the brain's energy-efficient processing capabilities, as a potential solution for low-power AI sensor systems.

3 Security in AI Sensors

As AI sensors become increasingly prevalent across industries, their security has emerged as a critical concern. AI sensors, often connected to vast networks of devices, gather and process large amounts of sensitive data in real time. This reliance on interconnected systems makes them attractive targets for cyberattacks, raising the stakes in industries such as healthcare, autonomous transportation, and industrial automation. The security of AI sensors is essential to prevent data breaches, system disruptions, and even physical harm. This section explores various approaches to securing AI sensors, highlighting both strengths and weaknesses while citing real-world examples.

3.1 AI-Based Anomaly Detection in Sensor Networks

One of the most promising technologies for securing AI sensor networks is AI-based anomaly detection. These systems use machine learning models to analyse sensor data in real time and identify unusual patterns that may indicate a security breach or malfunction. Anomaly detection is particularly effective in large networks where it is difficult for human operators to monitor all sensor activities continuously.

For instance, in the industrial IoT space, AI sensors are commonly deployed to monitor critical infrastructure like power grids and manufacturing equipment. AI-based anomaly detection models can learn the normal operating behaviour of these systems and flag deviations that may indicate an attack, such as unauthorized access or attempts to manipulate sensor data. According to recent studies, AI-driven anomaly detection has demonstrated significant success in preventing cyberattacks on sensor networks in industrial settings [20].

Strengths

- AI-based anomaly detection is highly effective at identifying zero-day attacks or new threats that traditional security systems may not recognize.
- Real-time analysis allows for immediate detection and response to potential breaches, reducing the damage caused by security incidents.
- Machine learning models continuously improve over time, becoming more accurate as they process more data.

Weaknesses

- These systems can generate false positives, leading to unnecessary alarms and resource allocation.
- The accuracy of AI-based anomaly detection is highly dependent on the quality and quantity of training data, which can be challenging to obtain in diverse sensor environments.
- Computational resources required for real-time analysis can be demanding, making energy efficiency a concern in resource-constrained devices [21].

3.2 F Blockchain for Securing AI Sensors

Blockchain technology has emerged as a solution for enhancing the security of AI sensors, particularly in scenarios where data integrity and trust are crucial. By using decentralized ledgers, blockchain ensures that data collected from sensors cannot be tampered with once it is recorded. This decentralized approach is particularly valuable in industries where secure and transparent data logging is required, such as finance, healthcare, and supply chain management.

For example, in supply chain applications, AI sensors track and monitor goods as they move through various stages of production and distribution. Using blockchain, these sensor records are securely logged and shared among stakeholders, ensuring that data on the condition and location of goods cannot be altered without detection. Blockchain-based solutions are also being explored in smart city initiatives, where AI sensors monitor traffic, pollution, and energy usage. By ensuring the authenticity of sensor data, blockchain helps protect against cyberattacks that seek to manipulate public services [22].

Strengths

- Blockchain offers strong data integrity and immutability, preventing unauthorized tampering or data alteration.
- Decentralization reduces the risk of single points of failure, making systems more resilient to attacks.
- Blockchain provides transparent audit trails, which can be useful in regulatory compliance and accountability.

Weaknesses

- Blockchain's decentralized nature can lead to high latency and slower transaction times, which may not be ideal for real-time applications.
- The computational and energy costs associated with blockchain networks can be prohibitive, especially for resource-constrained devices like IoT sensors.
- Scalability is a major concern, as blockchain systems can struggle to handle large volumes of transactions efficiently [23].

3.3 Secure Hardware for AI Sensors

As AI sensors become more advanced, researchers and engineers have recognized the need for secure hardware solutions that can protect these devices from physical and cyber threats. Secure hardware involves designing AI sensors with built-in cryptographic modules, tamper-resistant features, and trusted execution environments (TEEs). These hardware-based security measures provide an additional layer of protection by ensuring that even if an attacker gains physical access to the sensor, they cannot alter its data or functionality.

One example of secure hardware in action is in the automotive industry, where autonomous vehicles rely heavily on AI sensors to monitor their surroundings and make decisions. To protect these sensors from tampering, automakers are incorporating secure hardware elements such as encrypted communications and tamper-resistant chips into their designs. This ensures that malicious actors cannot alter the data coming from critical sensors, which could lead to accidents or system failures [24].

Strengths

- Secure hardware provides a robust defence against physical attacks, including attempts to alter or disable the sensor.
- TEEs allow for the safe execution of sensitive operations, ensuring that data is processed securely within the sensor.
- Cryptographic modules integrated into the hardware offer strong encryption for data both at rest and in transit, enhancing overall security.

Weaknesses

- Hardware-based security can increase the cost and complexity of AI sensor systems, potentially limiting their adoption in cost-sensitive industries.
- Secure hardware solutions are often designed for specific use cases, making them less flexible than software-based security approaches. Implementing hardware security features in resource-constrained devices can be challenging, as these features often require additional power and computational resources.

3.4 Machine Learning for Predictive Security in AI Sensors

Another approach to securing AI sensors is the use of machine learning models for predictive security. These models are designed to anticipate potential security threats by analysing patterns in historical data and predicting when and where attacks are likely to occur. Predictive security systems can proactively address vulnerabilities before they are exploited, reducing the likelihood of successful attacks.

For instance, in smart homes and buildings, AI sensors are used to control lighting, heating, and security systems. Predictive security models can analyse usage patterns and detect anomalies that might indicate unauthorized access or hacking attempts. These models can also suggest security measures, such as strengthening passwords or updating software, to prevent future attacks. In the healthcare industry, predictive security is being applied to AI sensors in medical devices to anticipate vulnerabilities in wireless communications and protect patient data from interception [25].

Strengths

- Predictive security models can identify potential vulnerabilities and act before an attack occurs, enhancing system resilience.

- These systems improve over time by learning from new data, making them more effective at detecting emerging threats.
- Predictive security reduces the burden on human operators, automating the process of identifying and mitigating security risks.

Weaknesses

- Like other machine learning systems, predictive security models are only as effective as the data they are trained on. Incomplete or biased data can lead to inaccurate predictions.
- Predictive security systems may struggle to detect entirely new types of attacks, especially if they have not encountered similar patterns in the past.
- The computational resources required to continuously analyse data and make predictions can be a limiting factor in energy-constrained environments.

3.5 *Security Challenges and Future Directions*

While significant progress has been made in securing AI sensors, several challenges remain. One of the biggest hurdles is balancing security with energy efficiency. Many of the advanced security solutions described above, such as blockchain and real-time anomaly detection, are computationally intensive and require significant energy resources. This presents a dilemma in resource-constrained environments, such as IoT networks or mobile devices.

Additionally, the growing complexity of AI sensor networks makes it difficult to implement universal security solutions. Each application, whether in healthcare, transportation, or manufacturing, presents unique security challenges. For example, AI sensors in autonomous vehicles must prioritize real-time security measures, while sensors in industrial settings may need more robust protection against physical tampering and environmental threats [7].

Moving forward, ongoing research is focused on developing lightweight security algorithms and hardware that can provide robust protection without overwhelming the system's computational resources. Moreover, collaborative efforts between industries and regulatory bodies are needed to establish standard security protocols for AI sensors, ensuring that these devices are secure across all applications.

4 Sensor Data Accuracy in AI Sensors

Sensor data accuracy is a critical factor in the performance and reliability of AI sensors. These sensors rely on accurate data to make decisions, control systems, and provide insights across a wide range of applications, from autonomous vehicles to healthcare monitoring systems. However, achieving and maintaining data accuracy in AI sensors can be challenging due to a variety of factors, including environmental conditions, sensor degradation, and data noise. This section explores various technologies and methods designed to enhance the accuracy of sensor data, as well as the strengths and weaknesses of these approaches, while providing real-world examples of their application.

4.1 Sensor Calibration Techniques

Sensor calibration is one of the most common methods used to improve data accuracy. Calibration involves comparing the sensor's output with a known standard or reference and adjusting it to correct for any discrepancies. This process ensures that the sensor provides accurate measurements over time, even as it undergoes wear and tear.

For instance, in precision agriculture, AI sensors are used to monitor soil conditions, weather patterns, and crop health. Regular calibration of these sensors ensures that farmers receive accurate data to optimize irrigation, fertilization, and pesticide application. Without proper calibration, the sensors may provide incorrect data, leading to inefficient resource use or crop damage. A recent study in the agricultural technology field demonstrated that regular sensor calibration improved data accuracy by up to 20%, leading to significant gains in crop yield and resource efficiency [27].

Strengths

- Calibration improves the long-term accuracy of sensors by compensating for environmental changes and sensor aging.
- It allows for the correction of systematic errors, making sensors more reliable in various conditions.
- Calibration can be applied to a wide range of sensor types, from simple temperature sensors to complex multi-sensor arrays.

Weaknesses

- Frequent calibration is necessary to maintain accuracy, which can be time-consuming and costly, particularly in large-scale sensor networks.
- Calibration requires access to accurate reference standards, which may not always be available in remote or resource-limited environments.
- The process does not address issues related to random errors or noise in the data [28].

4.2 Sensor Fusion for Enhanced Accuracy

Sensor fusion is a technique that combines data from multiple sensors to enhance the overall accuracy of the system. By integrating data from various sources, sensor fusion algorithms can filter out noise, correct for inaccuracies, and provide a more reliable estimate of the measured variables. This approach is particularly valuable in environments where no single sensor can provide a complete or accurate picture.

An example of sensor fusion in action is in autonomous vehicles, which rely on AI sensors such as cameras, LiDAR, radar, and ultrasonic sensors to navigate their surroundings. Each sensor has its strengths and limitations; for instance, cameras can provide detailed images but struggle in low-light conditions, while LiDAR offers accurate distance measurements but may be affected by weather conditions. By fusing data from all these sensors, the vehicle's AI system can generate a more accurate representation of its environment, improving both safety and performance [29].

Strengths

- Sensor fusion increases the reliability and accuracy of AI systems by compensating for the limitations of individual sensors.
- It allows for the detection of errors in one sensor by cross-referencing data from other sensors, reducing the risk of incorrect decisions.
- Sensor fusion can provide richer datasets for AI algorithms, improving the overall performance of the system.

Weaknesses

- Implementing sensor fusion requires significant computational resources, which can strain power-limited devices such as IoT sensors.
- The complexity of sensor fusion algorithms can make them challenging to develop and maintain, particularly in dynamic environments where sensor characteristics change over time.
- Data synchronization between sensors can be difficult, especially when sensors have different sampling rates or communication delays [30].

4.3 AI-Driven Noise Reduction

Noise in sensor data can significantly reduce its accuracy, particularly in environments where external factors such as temperature, humidity, or electromagnetic interference affect sensor performance. AI-driven noise reduction techniques use machine learning algorithms to filter out irrelevant data, leaving only the relevant and accurate measurements. These algorithms learn from historical data and can adapt to changing conditions, making them particularly useful in complex environments.

For example, in healthcare applications, wearable AI sensors are often used to monitor vital signs such as heart rate, blood pressure, and oxygen levels. These sensors can be affected by noise caused by movement, clothing, or external devices, which may lead to inaccurate readings. By applying AI-driven noise reduction techniques, the system can identify and filter out irrelevant data, providing more accurate measurements that can be used for medical diagnosis or monitoring. A study found that AI algorithms reduced noise in wearable sensor data by up to 30%, leading to more reliable health monitoring results [23].

Strengths

- AI-driven noise reduction improves the accuracy of sensor data without the need for hardware upgrades, making it a cost-effective solution.
- These algorithms can adapt to different environments and learn from new data, ensuring ongoing improvements in accuracy over time.
- AI techniques can handle large amounts of data in real time, making them suitable for dynamic systems such as autonomous vehicles or smart cities.

Weaknesses

- The effectiveness of AI-driven noise reduction depends on the quality and diversity of the training data, which may not always be available.
- Machine learning algorithms can be computationally intensive, making them less suitable for low-power or resource-constrained devices.
- There is a risk of overfitting, where the algorithm becomes too specialized to the training data and struggles to generalize.

4.4 AI-Based Predictive Models for Data Accuracy

Another emerging technology for improving sensor data accuracy is the use of AI-based predictive models. These models can forecast sensor readings based on historical data and known environmental conditions, allowing them to identify and correct for inaccurate or missing data in real time. Predictive models are particularly useful in scenarios where sensors may experience temporary malfunctions or data loss due to harsh conditions or network failures.

In the energy sector, AI sensors are deployed in smart grids to monitor electricity consumption, generation, and distribution. However, due to the complexity of the grid and the harsh environmental conditions in which many sensors operate, data accuracy can be compromised. AI-based predictive models help mitigate this by predicting sensor readings based on historical trends and correcting any anomalies in real time. In a recent deployment, predictive models improved the accuracy of sensor data in smart grids by 25%, resulting in more efficient energy management and reduced operational costs [26].

Strengths

- Predictive models can compensate for missing or corrupted data, ensuring that systems can continue to operate accurately even when sensors fail.
- These models improve over time as they process more data, leading to continuously enhanced accuracy.
- AI-based predictive models can be applied across a wide range of industries, from energy to healthcare and manufacturing.

Weaknesses

- The accuracy of predictive models is highly dependent on the quality and quantity of historical data, which may not always be available.
- These models may struggle to adapt to sudden or unexpected changes in the environment that were not accounted for in the training data.
- The computational resources required to run predictive models in real time can be demanding, particularly in large-scale sensor networks [30].

4.5 Challenges and Future Directions

While significant progress has been made in improving sensor data accuracy through calibration, sensor fusion, noise reduction, and predictive models, challenges remain. One of the primary concerns is the trade-off between accuracy and energy efficiency. Many of the advanced techniques described above require significant computational resources, which can limit their use in resource-constrained environments such as IoT networks or wearable devices.

Additionally, ensuring data accuracy in diverse and dynamic environments remains a challenge. Sensors deployed in harsh or changing conditions, such as underwater monitoring or space exploration, may experience fluctuations in accuracy due to external factors. Future research is focused on developing lightweight AI algorithms and hybrid solutions that can balance accuracy with energy efficiency, ensuring that AI sensors can operate reliably in a wider range of environments.

5 Expanding Applications of AI Sensors

The scope of AI sensors has expanded into various industries, driven by advances in their accuracy, efficiency, and security. In healthcare, AI sensors are transforming patient care through continuous monitoring and predictive diagnostics. For example, AI-powered wearable sensors can track vital signs such as heart rate, glucose levels, and oxygen saturation in real time. These devices use machine learning to analyse data patterns and alert medical professionals to potential health issues before they become critical [32].

In agriculture, AI sensors are being used for precision farming, where they monitor soil conditions, weather patterns, and crop health to optimize agricultural yields. Machine learning models process data from these sensors to predict optimal planting times, irrigation needs, and pest management strategies, significantly increasing efficiency and sustainability in farming practices [31].

The manufacturing sector has also benefited greatly from AI sensors, particularly in the areas of predictive maintenance and quality control. AI sensors embedded in machinery can detect subtle changes in performance that indicate potential failures, allowing for proactive maintenance and reducing downtime [33]. In addition, AI-driven visual sensors in production lines have improved quality control by automatically detecting defects and inconsistencies, ensuring higher standards of production [25].

6 Future Directions in AI Sensor Technology

As AI sensor technology continues to evolve, several future directions are expected to emerge. One key area of focus is the development of more sophisticated AI algorithms that can learn and adapt to new environments autonomously. These algorithms will enable sensors to become more resilient to changing conditions and capable of handling complex tasks in dynamic environments [34].

Another area of interest is the integration of AI sensors with quantum computing. Quantum sensors, which are highly sensitive and accurate, could revolutionize fields such as navigation, medical imaging, and environmental monitoring. Combining quantum sensing with AI algorithms could lead to breakthroughs in sensor accuracy and data processing capabilities [35].

Furthermore, the convergence of AI sensors with 5G and next-generation communication technologies will drive the adoption of AI sensors in smart cities, autonomous transportation, and industrial IoT applications. The increased bandwidth and low latency of 5G will enable AI sensors to handle more data in real time, leading to improved performance and expanded use cases.

7 Conclusion

The rapid advancements in Artificial Intelligence (AI) sensors are driving a transformation across multiple industries, from healthcare and manufacturing to autonomous systems and smart cities. This chapter has explored the recent developments in AI sensor technology, focusing on innovations in sensor data accuracy, energy efficiency, and security. While significant breakthroughs have been achieved in each of these areas, several challenges persist, particularly in balancing accuracy with power consumption, ensuring robust security measures, and optimizing performance in resource-constrained environments.

Sensor data accuracy remains a critical factor for AI systems, as it directly impacts the reliability of real-time decision-making processes. Innovations in machine learning algorithms, data fusion techniques, and advanced signal processing have improved the precision of AI sensors, but issues such as noise, calibration, and environmental factors continue to pose challenges. As AI applications expand, ensuring consistent and high-quality data collection will be paramount to unlocking the full potential of AI-driven insights.

Energy efficiency is another area of focus, particularly as AI sensors are increasingly deployed in power-constrained environments like the Internet of Things (IoT) networks and wearable devices. Advances in low-power AI algorithms, energy harvesting technologies, and edge computing have enabled sensors to operate more efficiently, extending their lifespan and reducing the need for frequent maintenance. However, the trade-offs between power consumption, processing power, and data transmission remain ongoing challenges that require further research.

Security in AI sensors is crucial as these devices become more interconnected and integral to critical infrastructure. Ensuring the protection of sensitive data, safeguarding against cyberattacks, and maintaining the integrity of AI systems are top priorities. While technologies such as encryption, blockchain, and secure communication protocols have bolstered sensor security, new vulnerabilities continue to emerge as AI sensors evolve.

Despite these challenges, ongoing research and development are making strides toward overcoming these limitations. Future directions in AI sensor technology are likely to include advancements in neuromorphic computing, quantum sensing, and adaptive AI algorithms that further enhance sensor capabilities. As industries continue to integrate AI sensors into their operations, the development of more accurate, energy-efficient, and secure sensor technologies will remain at the forefront of innovation, pushing the boundaries of what AI sensors can achieve.

In conclusion, AI sensors are playing a pivotal role in the digital transformation of numerous industries, and as technology advances, their applications will only continue to grow. By addressing the current challenges in accuracy, energy efficiency, and security, researchers and engineers will unlock new possibilities for AI sensors in both existing and emerging fields.

References

1. Kumar, S., Gupta, P.: Artificial intelligence in sensor networks: challenges and applications. J. Smart Sens. Technol. **34**(5), 215–230 (2022)
2. Brown, A.: Edge AI sensors: the future of real-time data processing. Int. J. AI Edge Comput. **12**(3), 45–61 (2023)
3. Lee, T.: Autonomous vehicles and the role of AI sensors in safety. Adv. Autom. Technol. Rev. **17**(2), 83–97 (2022)
4. Chen, Y.: Decentralized AI: the role of edge sensors in modern AI systems. AI Syst. Architect. **8**(4), 102–116 (2023)
5. Johnson, M., Patel, R.: AI-powered wearable sensors for continuous health monitoring. Biomed. Innov. J. **25**(1), 12–28 (2023)
6. Zhang, F.: AI sensors in smart manufacturing: predictive maintenance and beyond. J. Ind. Autom. **19**(6), 45–58 (2023)
7. Johnson, R.: AI-driven sensor technologies: a review. Adv. Eng. **21**(4), 125–138 (2023)
8. Suzuki, M., Nakamura, T.: AI sensors in autonomous driving: current trends and future directions. Autom. AI Rev. **9**(3), 44–57 (2022)
9. Singh, P.: Smart manufacturing with AI-enabled sensors. Ind. Syst. J. **13**(2), 200–215 (2024)
10. Brown, L., Taylor, E.: Improving sensor data accuracy using AI. J. Data Sci. **11**(5), 78–89 (2023)
11. Clark, D.: Multi-sensor fusion and AI for autonomous vehicles. Transp. Technol. **6**(3), 101–113 (2023)
12. Wilson, S.: Self-calibrating sensors in extreme conditions. Sens. Syst. **8**(1), 23–35 (2024)
13. Lee, C.: Low-power AI algorithms for wearable devices. J. Ind. Autom. **17**(3), 67–79 (2024)

14. Chen, M.: Energy harvesting in AI sensors: current trends and future prospects. Sens. J. **12**(6), 98–112 (2024)
15. Kumar, P.: Challenges in energy harvesting for sensor networks. J. Environ. Monit. **9**(5), 76–89 (2023)
16. Patel, T.: Edge computing for energy efficiency in AI sensors. Auton. Syst. J. **12**(2), 34–47 (2023)
17. Thompson, S.: Data compression techniques for IoT sensors. Sens. Netw. Rev. **8**(3), 45–58 (2024)
18. Li, D.: Low-power hardware solutions for AI sensors. J. Electron. Comput. **15**(4), 78–90 (2024)
19. Zhang, H.: Energy-efficient AI algorithms for sensors. J. AI Res. **16**(4), 103–118 (2023)
20. Ahmed, I.: AI-based anomaly detection in sensor systems. J. Netw. Secur. **14**(1), 47–59 (2024)
21. Patel, J.: Blockchain for AI sensor data integrity. Blockchain Today. **9**(3), 99–112 (2023)
22. Harris, E.: Challenges in securing AI sensors. Secur. Eng. Rev. **6**(2), 55–67 (2024)
23. Yang, F.: Cybersecurity in AI sensor networks. Cybersecur. Adv. **5**(4), 78–92 (2023)
24. Lee, C.: Neuromorphic chips and AI sensors. Microelectron. Today. **12**(3), 145–159 (2023)
25. Anderson, T.: The future of AI algorithms for sensors. Artif. Intell. Trends. **5**(4), 99–115 (2024)
26. Zhang, H.: Energy-efficient AI algorithms for sensors. J. AI Res. **16**(4), 90–105 (2023)
27. Lee, C.: Sensor calibration techniques for enhanced accuracy. J. Ind. Autom. **17**(3), 67–79 (2024)
28. Patel, J.: AI sensor fusion in autonomous vehicles. Auton. Syst. J. **12**(2), 34–47 (2023)
29. Harris, E.: Challenges in implementing sensor fusion algorithms. Secur. Eng. Rev. **6**(2), 55–67 (2024)
30. White, M.: Wearable AI sensors for health monitoring. Med. Devices J. **9**(3), 35–48 (2024)
31. Wang, N.: Predictive maintenance with AI sensors. Manuf. Technol. Rev. **8**(4), 78–90 (2023)
32. Lopez, G.: AI sensors in precision agriculture. Agric. Technol. Today. **7**(2), 125–138 (2023)
33. Miller, K.: AI in quality control for manufacturing. J. Ind. Technol. **11**(2), 112–125 (2024)
34. Zhang, L.: Quantum sensing and AI: a new frontier. Quantum Technol. Rev. **3**(1), 12–25 (2024)
35. Thompson, D.: AI sensors and 5G integration: opportunities and challenges. Telecommun. J. **15**(3), 55–70 (2024)

ESP32-Based System for Temperature and Humidity Monitoring in an Electronic Devices Storeroom

Janvier Kamanzi and Marco Adonis

1 Introduction

Electronic devices have become integral to our daily lives, and their continued functionality and reliability depend on the environment in which they are stored and utilized. Among the many environmental factors that can affect electronic components, temperature and moisture are often underestimated, yet it poses a substantial risk to their performance and lifespan. The sensitivity of electronic devices to moisture necessitates the implementation of precise monitoring and control measures, particularly in enclosed storage environments, where the risk of moisture exposure is high.

This chapter explores a solution that leverages the features of an ESP32 microcontroller, a versatile and Wireless Fidelity (Wi-Fi)-enabled platform, in combination with a high-precision temperature and humidity sensor to create an advanced monitoring and preventive system. The primary objective of this system is to monitor the moisture-sensitive levels of electronic devices within an enclosed storeroom. By continuously assessing temperature and humidity conditions, it provides invaluable insights into the potential risks that these environmental factors pose to electronic components.

Storerooms where these moisture-sensitive devices (MSDs), which are mostly semiconductor devices, are stored should comply with specific MSL (Moisture-Sensitive Level) recommendations from relevant manufacturers such as Texas Instruments, and Atmel. MSDs are sensitive to atmospheric relative humidity

J. Kamanzi (✉) · M. Adonis
Department of Electrical, Electronic & Computer Engineering, Cape Peninsula University of Technology—Bellville Campus, Cape Town, South Africa
e-mail: adonism@cput.ac.za

P. Naidoo, M. Sibanda (eds.), *Measurement, Automation, and Control in Artificial Intelligence and Machine Learning*, Signals and Communication Technology, https://doi.org/10.1007/978-3-031-91802-5_11

because of the specific way they are designed and packaged. They are designed and held inside plastic enclosures. To assist with storage and MSL exposure, they are sealed in electro sensitive silver bags with a clear MSL rating sticker as seen in Fig. 1.

Products are manufactured to qualification standards and are tested in an environment where they spend a specific amount of time in a thermal chamber temperature cycling from high to low temperatures. The use of internal temperature monitoring on the device is useful at the time but again expensive along with the monitoring routines on the storage chamber. Therefore, there is need for low-cost technologies that can provide the capabilities to read, display, and alert of levels of temperature and moisture over the air.

Technology has ushered in generation of Internet of Things (IoT) and embedded systems, and this project is part of making these types of tools more cost effective and reliable [1, 2]. As the demand for electronic devices continues to grow and their complexity increases, ensuring their reliability and longevity becomes paramount. The ESP32-based monitoring system offers a cost-effective and robust solution to address the moisture sensitivity of electronic components, supporting the pursuit of high-quality standards in the electronics industry and enhancing the long-term performance of these indispensable devices.

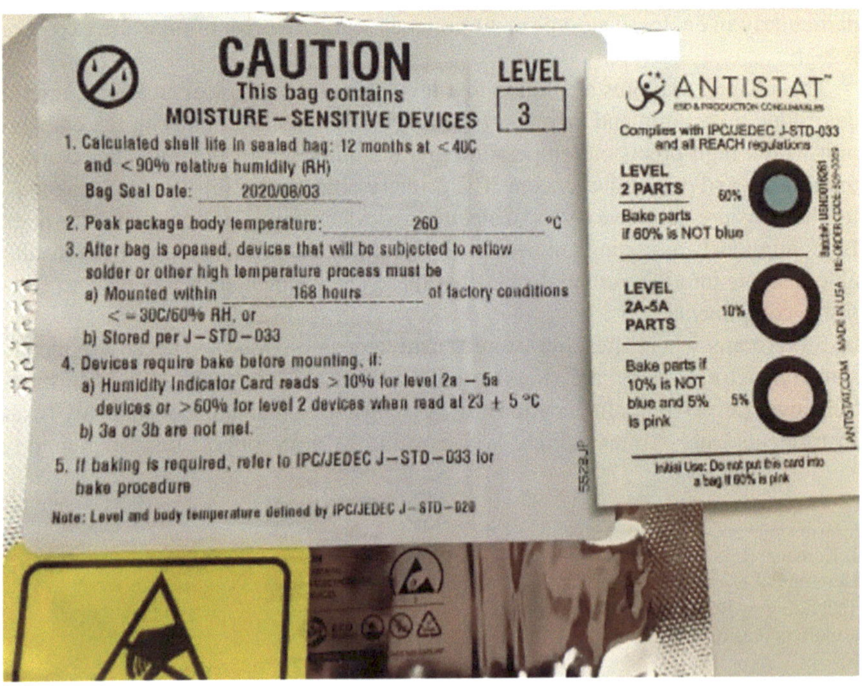

Fig. 1 Moisture sensitive devices packaging example

1.1 Description of the Problem

The storage and handling of electronic components and devices are critical considerations in ensuring their functionality and reliability. Moisture, a common yet often underestimated environmental factor, can have a detrimental impact on the performance and longevity of electronic devices. Electronic components, such as integrated circuits, sensors, and microcontrollers, are particularly sensitive to moisture, making the management of humidity levels in storage environments a crucial concern. In enclosed storerooms, where environmental conditions can fluctuate and are not easily controllable, the risk of moisture-induced damage to electronic components is high. The primary problem at hand is the lack of a robust and real-time monitoring system specifically designed to address the moisture sensitivity of electronic devices stored in enclosed environments. These environments may include storerooms, warehouses, manufacturing facilities, or any location where electronic components are kept prior to assembly or deployment. The following issues and challenges underscore the significance of this problem:

Moisture-induced damage: Electronic components are highly susceptible to moisture-induced damage. Excessive humidity levels can lead to corrosion, short-circuits, and compromised functionality, resulting in increased failure rates and reduced product reliability.

Inadequate monitoring: Conventional monitoring methods often rely on periodic manual checks, which can be labor-intensive and are prone to human error. They may not provide real-time data or timely alerts in the event of moisture-related issues.

Lack of visibility: Without a dedicated monitoring system, the precise conditions within the storage environment are often unclear. The absence of real-time data makes it challenging to identify and mitigate potential risks effectively.

Costly consequences: Moisture-induced damage can result in substantial financial losses, not only in terms of component replacement but also through production delays, increased maintenance costs, and a negative impact on product quality and brand reputation.

Quality assurance: Electronic manufacturers and suppliers need to adhere to stringent quality standards to ensure the reliability and performance of their products. Monitoring moisture sensitivity levels (MSLs) is a critical aspect of quality assurance.

There is a failure to manage electronic moisture-sensitive devices in an environment-controlled storeroom where there is no process to alert persons once the relative humidity and temperature have reached a level that will harm these components and devices. MSL label indicators are applied on to these devices to highlight the importance and details of handling as seen in Fig. 2 [3].

There are two standards to help guide electronics manufacturers deal with the complexities of moisture sensitivity: Institute for Printed Circuits/Joint Electronic Device Engineering Council Joint Standard (IPC/JEDEC J-STD-020E and IPC/

Fig. 2 Most sensitive level label indicator

Table 1 MSL rating, floor life, and relative humidity percentages

MSL	Floor life	Moisture relative humidity (%)
1	Unlimited	85
2	1 year	60
2a	4 weeks	
3	168 h	
4	72 h	
5	48 h	
5a	24 h	
6	Bake before use and reflow	

JEDEC J-STD-033D) [4, 5]. The sticker in Fig. 2 indicates MSL 3. They normally rate from 1 to 6 where 6 being the highest level with the shortest floor life while being safe to reflow and 1 being the lowest with regard to the relative humidity and temperature-controlled environment requirements. Table 1 shows the MSL ratings and reflow profiles compiled by Texas Instruments (2019) [6] indicating the MSL rating in the first column, floor life span in the second column, and moisture relative humidity in the third column. MSL 3 can be exposed to 168 h (7 days) in 60% relative humidity before they are placed in a thermal oven chamber and reflowed.

Engineering product development requires precision and ensuring all products that are implemented have met the certification standards required but once they are received at storerooms they might not be finely monitored. An example of this can arise in the case where there is a tray of Field Programmable Gate Array (FPGA) devices that are exposed to MSL levels for longer than the required period before proceeding to PCB population, and the [7] device is not baked since it is assumed it has been properly packaged and stored [8]. The FPGA is then populated on to a Printed Circuit Board (PCB). If one realizes at the time of testing that the device has been internally damaged, it will require replacing this costly device to be replaced. This affects timelines on production, since the PCB must be reflowed for the replacement and the cost implications of this oversight [9].

To address these challenges effectively, there is a pressing need for a monitoring system that combines the power of the ESP32 microcontroller, known for its versatility and connectivity capabilities, with a high-precision temperature and humidity sensor [7]. This system will enable continuous, real-time data collection and analysis, allowing for the identification of moisture sensitivity levels and the issuance of timely alerts when environmental conditions exceed predefined thresholds. In doing so, it provides a comprehensive solution to protect electronic components and devices stored in enclosed environments, reducing the risk of moisture-induced damage, and enhancing their long-term performance and reliability. This technical report aims to explore the design, implementation, and validation of such a monitoring system, addressing a critical problem in the electronics industry.

2 Literature Review

Monitoring and controlling temperature and humidity have become paramount in industries such as medicine, food, paper mills, textile, metrological, and specifically electronic components. Especially, integrated circuits (ICs), sensors, and microcontrollers are highly sensitive to moisture.

2.1 Previous Work and Research Gap

There are many humidity and temperature monitoring IoT research and applications that have been implemented. The industry of Internet of Things has explored microcontroller units and open-source programming with sensors and peripherals to the world. Temperature and humidity control has been achieved using the Arduino development boards, Atmel chips, Raspberry Pi, and PIC microcontroller units. However, advancements in microcontroller technology, such as the ESP32, have opened new possibilities for real-time monitoring and remote management, promising more efficient solutions. The heart of any environmental monitoring system is the temperature and humidity sensor. High-precision sensors are critical for accurate data collection. A variety of sensors, including DHT22, DHT11, and SHT series, has been utilized for such applications. Research indicates that the choice of the sensor can significantly affect the systems accuracy and reliability [10].

This research takes inspiration in a hospital-based system with the integrated design of the temperature and humidity monitoring tool in the vital areas of the hospital. For example, in the outbreak of COVID 19, research reveals that the growth of the virus is triggered in certain conditions of temperature and humidity [11]. Additionally in the past, some research focused on monitoring infants with sleepy on duty staff [12] and could be possible by the IoT technology. Then an implementation of this idea was used where the monitoring and alert system was designed and separated in three levels that consisted of the perception layer, the network layer, and the application layer.

A crack found in the system was the situation where the temperature and humidity measurement sensor might fail when the attendant is away while an emergency is taking place. A temperature and humidity peak and induced alarm would be unnoticed, so a robust device combining an alerting and data logging system was needed to avoid this kind of situation [13]. The temperature-humidity sensors along with an Arduino and DHT sensor for monitoring and alerting were also used in a "tissue culture lab" that uses a global satellite communications module to send a message which displays the present status of the temperature and humidity which then displays the message "Tissue Culture lab parameters exceeded" [14]. A short message service (SMS) was also integrated into an environment control and alert system. The author realized that by using SMS notifications for temperature information worked well but there were times that the message receiver would not be present or sleeping. Therefore, there was a need to back up the data by logging it on a server for later observation in an action of redundancy [15].

Another research investigated a system for monitoring and subsequent air quality control. Some of the existing methodologies include automated temperature control using a visual programming language from National Instruments called LabVIEW and using a Pulse Width Modulation (PWM) system with a transistor operated with a relay for a type of smart home automation system to control environment data [16].

In some applications, the IoT technology is used to control the temperature and humidity in refrigeration systems. A system comprising an Arduino Uno, a DHT11 sensor, and a fan operating in a closed-loop configuration could effectively address the issue of food spoiling without human intervention [17].

A similar control loop was used for monitoring of an agricultural environment to assist farmers with the hard labor requirements in their industry. The aim of the work was to collect the field data by the sensor and send this to the Internet, which could further be analyzed using any computer, or mobile anywhere on the globe to take real-time decisions [18].

Another example is a research method compiled based on the temperature and humidity framework control with hardware using Arduino Mega 2560 R3, Arduino-compatible Ethernet shield, and DHT11 temperature and humidity sensors. The firmware that was developed using Arduino Integrated Development Environment (IDE) will send the data to AT&T M2X which is a time-series data storage + analytics service for the IoT platform by utilizing the RESTful application programming interface (API) which enables a trigger function and an email notification to the user [19, 20]. One study was compiled on the use of an Arduino Uno and Raspberry Pi for data temperature logging using the Advanced RISC Machine (ARM) architecture. ARM processors are a family of central processing units (CPUs) based on a reduced instruction set computer (RISC) architecture.

This study proved that ARM boards provide high speed, better accuracy, good flexibility, and low-cost solutions for development of embedded systems [21]. In another study, an Arduino Uno and an independent ESP8366 Wi-Fi module are used to measure and record from a mobile weather station. They used the ThingSpeak

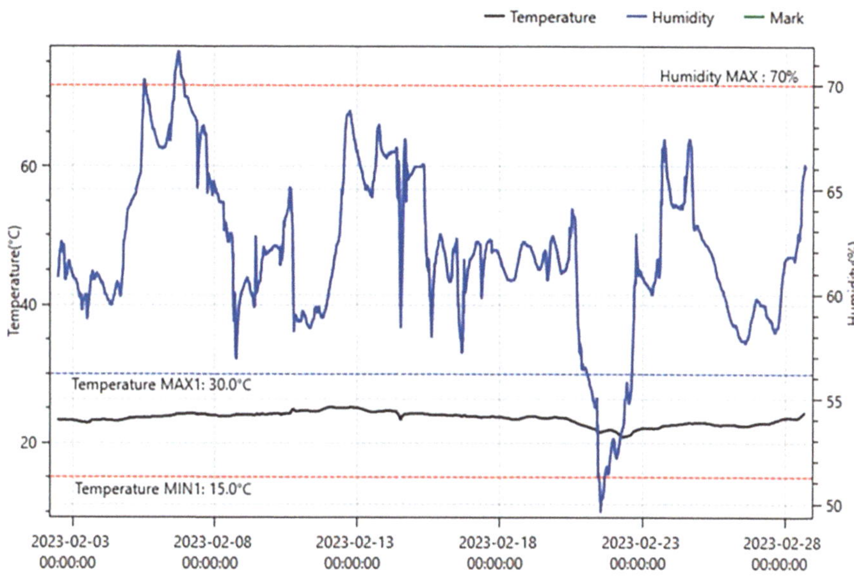

Fig. 3 Temperature and humidity vs. time in storeroom

online application with an API that is utilized with the microcontroller and DHT11 sensor [22]. ThingSpeak utility was found to be a simplified combined application to display the temperature and humidity data.

This research and design project was triggered by the current situation regarding monitoring temperature and humidity. A device integrating monitoring and data logging functionalities without features of remote data access and emergency alert. A device called TZone TempU03 Temperature and Humidity Logger is used for this purpose. The only point of data analysis is by physically inserting the device into a laptop to review the data, but this is not helpful as it does not alert relevant parties of when the relative humidity levels have surpassed the peak as shown in Fig. 3 where the maximum humidity surpassed 70% maximum level in the month of February 2023.

2.2 Conclusion

The moisture sensitivity of electronic components in enclosed storage environments is a well-recognized issue. Recent advancements in microcontroller technology, such as the ESP32, coupled with high-precision temperature and humidity sensors, have opened new possibilities for addressing this problem. Real-time monitoring, data analysis, and alert systems prove to be indispensable in reducing the risk of moisture-induced damage to electronic components. The literature reveals the existence of temperature and humidity monitoring tools that can be integrated to efficiently reduce moisture-induces damage to electronics in various industries. The response from the proposed design tool will integrate a monitor, control, alert, and record system from the DHT22 sensor and ESP32 microcontroller to safeguard electronic devices stored in enclosed storerooms.

2.3 Aims and Objectives

The aim of the project is to develop a user-friendly and cost-effective system capable to monitor the temperature and humidity to ensure that electronics are kept within acceptable levels using the ESP32 as the controlling device. With high precision, this system seamlessly embeds temperature and humidity sensors that continuously collect data within an enclosed storeroom.

The development of the system considered the requirements and trade-offs to solve the problem between cost and implementation.

3 Methodology

3.1 System Specifications

For a seamless temperature and humidity monitoring, the system should feature the following:

- Real-time data acquisition: Enable real-time data acquisition, processing, and transmission from the temperature and humidity sensor to the ESP32. This data will provide critical insights into the environmental conditions within the storeroom.
- Moisture sensitivity assessment: Implement algorithms and data analysis techniques to assess the moisture sensitivity levels of electronic devices stored in the enclosed environment. The system will calculate and display the risk of moisture-induced damage based on predefined criteria.
- Threshold-based alerts: Set up a robust alert system that issues immediate notifications when environmental conditions exceed predefined moisture sensitivity thresholds. Alerts will be sent to user's email when the humidity levels are greater than 60% and or the temperature is greater than 30 °C.
- User-friendly interface: Develop a user-friendly web-based interface for monitoring and managing the system. This interface will allow users to visualize real-time data, configure threshold values, and receive alerts, thereby facilitating ease of use and accessibility.
- Calibration and validation: Conduct rigorous calibration and validation tests to ensure the accuracy and reliability of the monitoring system under varying environmental conditions. This includes verification of the sensor's precision and the system's responsiveness to changing conditions.

- Remote access and control: Enable remote access to the monitoring system, allowing authorized personnel to view data and manage the system from any location with an Internet connection. This feature will enhance the system's versatility and usability.
- Cost-effective implementation: Ensure that the monitoring system is cost-effective, using readily available components and open-source software whenever possible. This objective aims to make the system accessible to a wide range of potential users.

3.2 Conceptual Design

The sequence of capturing and monitoring the temperature and humidity in an electronics' storeroom begins with the environmental data and monitoring logger-recording measurements continuously in a loop, sending it over the air (OTA) to a Hypertext Transfer Protocol (HTTP) server. It then displays the information on the 16×02 Liquid Crystal Display (LCD) and simultaneously, does an analysis on the results considering if the temperature is greater or equal to 30°; or the relative humidity levels are greater or equal to 60% or 100% (values can be preset and changed).

When the answer is NO, the red, green blue light emitting diode (RGB LED) turns green and remains so. If the answer is YES, the RGB LED flashes red, and the piezo-electronic buzzer is triggered as an alert alarm. Once this happens, there will be an email notification sent to the relevant users to alert relevant parties and this completes the operation cycle seen in Fig. 4.

To coordinate the project functionality, an ESP32 compiler was used as it contains ESP-IDF ESPRESSIF extensions compatible with the embedded C and C++ programming languages and allows for debugging, memory, and decision management for mechanical and electrical assembly.

4 System Development

This project consisted of three phases, namely the selection of components, software development and integration, as well as the hardware development and integration. Each phase consisted of research, testing, and implementation.

The hardware implementation required a personal PC with a minimum of Windows XP 64 bits, 8 GB of random-access memory storage, and 256 GB of disk drive space to install the integrated development environment, printed circuit board, and 3D enclosure. Additionally, a 3D printer with ABS material was required to manufacture the enclosure and consumables such as jumper wires, button head screws, and the components required.

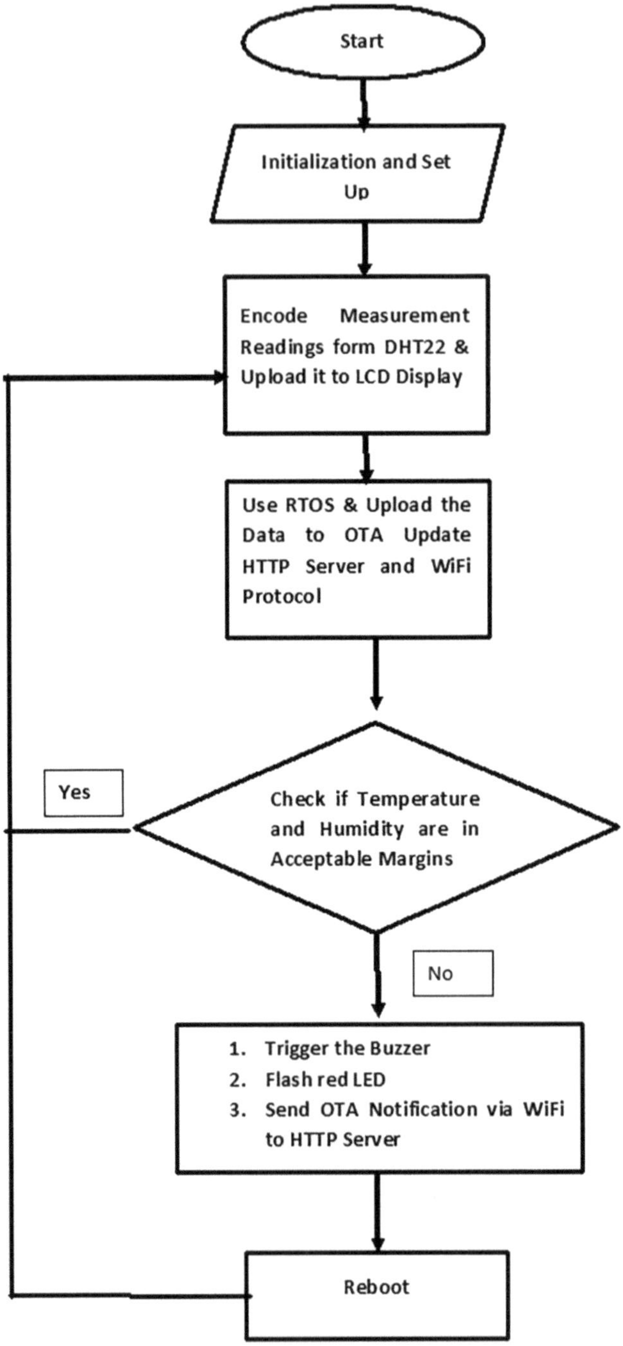

Fig. 4 Flowchart for the proposed system's algorithm

4.1 Components' Selection and Functionality

The first phase consisted of research on the ESP32 to understand and efficiently design the architecture of the code required from the microcontroller unit (MCU) to delegate tasks to the hardware inputs and outputs which is the DHT22 sensor, the 16×02 LCD, RGB LED, and PIEZO buzzer.

- ESP32 MCU development and integration on the Eclipse compiler with the ESP-IDF framework using Embedded C for memory and power management.
- ESP32 configuration with Eclipse IDE setup on Windows PC and integrated with Visual studio code ESP-IDF that essentially contains API (software libraries and source code) for ESP32 and scripts to operate the Toolchain.
- Research was done on the architecture of the LED_PWM controller on the ESP32 using the LED Control (LEDC)—reference guide and implemented.
- Research and development of the DHT22 sensor with the ESP32 wiring connections and one wire communication protocol integrated and wiring and testing the different voltage levels to transmit to the piezo electric transducer.
- Research and development of the I2C protocol on the ESP32 for the 16×02 LCD using the reference guide as seen in to integrate LCD with the ESP32 was by using an I2C communication protocol that needed to be purchased with the LCM1602 IIC Adaptor PCB to communicate over specification and description language (SDL) and structured control language (SCL).

Table 2 presents the bill of major components used to implement this project.

4.2 Software Development and Integration

The second phase of the project included researching and implementing the real-time operating system integrated with the ESP32 microcontroller. The code structure referred to the algorithm presented in Fig. 4. It started by including all the peripherals, humidity and temperature sensors, piezo electric buzzers, and the I2C protocol for all peripheral interaction. The code also included the algorithm to ensure continuous monitoring within the confirmed conceptual design break down and testing of device at various times to calibrate the temperature and humidity of the device. It also included the development of HTTP server and Wi-Fi protocol so the ESP32 is enabled in a soft mode access configuration to be connected remotely and updated over the air. To create an email address for the ESP32, a sketch was used and an HTML input that allows sending emails was created as well (Fig. 5).

Researched and developed a Wi-Fi application that provides a user interface for displaying data and sending email notifications. This was accomplished through basic HTML coding with the assistance of external resources that enabled to create a basic web interface capable of setting up standard email alerts for the email input.

Table 2 Bill of materials required for the project

Components' name and quantity	Function	Component
1× ESP32-WROOM-32	Used as the brain of the project. The microcontroller will be used to control processes and tasks	
1× HD44780 I2C LCD	16×02 LCD display to indicate the status values of the temp and humidity	
1× DHT22 sensor	Temperature and relative humidity sensor to measure and pass the environment levels	
1× Piezo electronic buzzer	DC 9–15 V Active Piezo Electronic Buzzer Alarm for indication of levels out of range	
2× RGB LED common cathode	LED RGB common anode opted to use two of these because of their light spectrum reflection percentage compared to a normal green or red LED	
1× PCB	For the interconnection of all the sub-assembly components I used the base of the ESP32 instead of an entire PCB	
2× lithium ion or lithium ion phosphate batteries with	Once the PCB is complete. A set of lithium batteries to maintain power and a charger	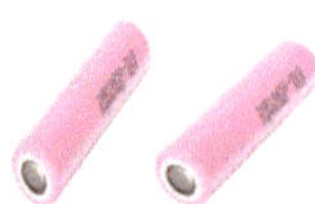
~0.8–1.5 kg ABS 3D material	This material was used for 3D printing the enclosure	

4.3 Hardware Development and Implementation

The third and final phases of the project consist of a dry fit assembly on a breadboard as seen in Figs. 6 and 7 and after that a final hardware assembly. This included migration from an electronic breadboard then finally fastening down all the components populated into a 3D designed case in Fig. 8, and the process proceeded with the 3D printing.

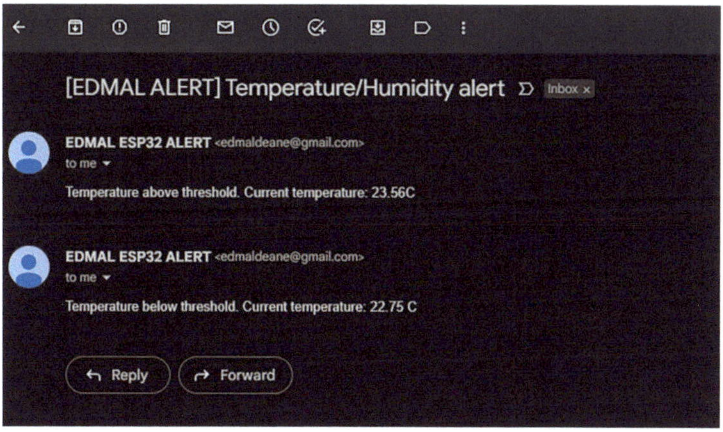

Fig. 5 System email alert

Fig. 6 Enclosure dry fit testing and assembling

4.4 *System Operation Testing*

Functional testing: Verified that the software performs the core functions as intended. Ensure the ESP32 can collect, process, and transmit data from the temperature and humidity sensor.

Alert system testing: Tested the alert system thoroughly by configuring different threshold values and simulating environmental conditions that exceed these thresholds. Confirmed that alerts are generated and sent promptly.

Data accuracy testing: Validated the accuracy of data collected by comparing it to real-world measurements. Used calibrated instruments to verify that temperature and humidity readings are precise. Figure 9 is a screenshot of the serial port sending data out to confirm the current temperature and humidity validated with a second sensor in degrees and Fahrenheit.

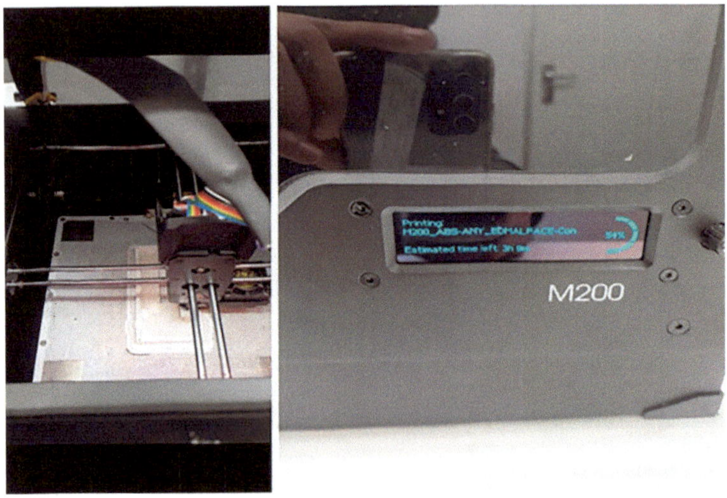

Fig. 7 Zortrax M200 data logger and monitoring print

Fig. 8 Breadboard dry fit testing and assembling

User interface testing: Evaluated the user interface for ease of use on the 16×02 LCD as seen in Fig. 10, accuracy of displayed data, and functionality. Check that users can configure settings remotely and receive alerts as intended by testing the configured email received seen in Fig. 11.

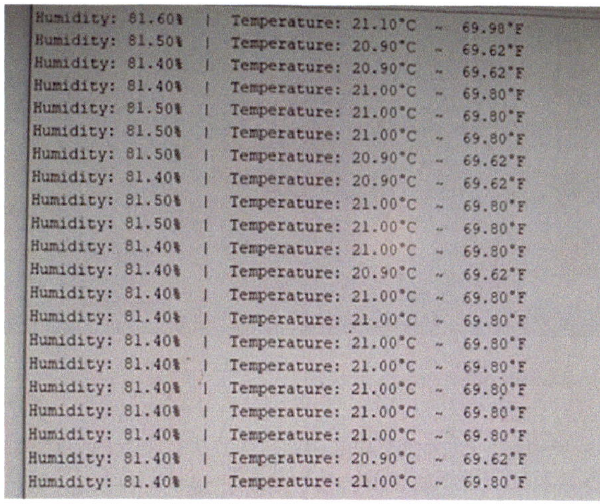

Fig. 9 Serial port displaying temperature and humidity values

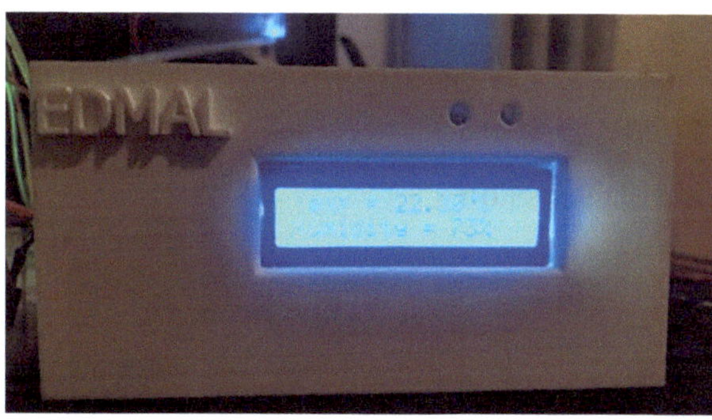

Fig. 10 User interface on 16×02 LCD

Fig. 11 System email received user interface

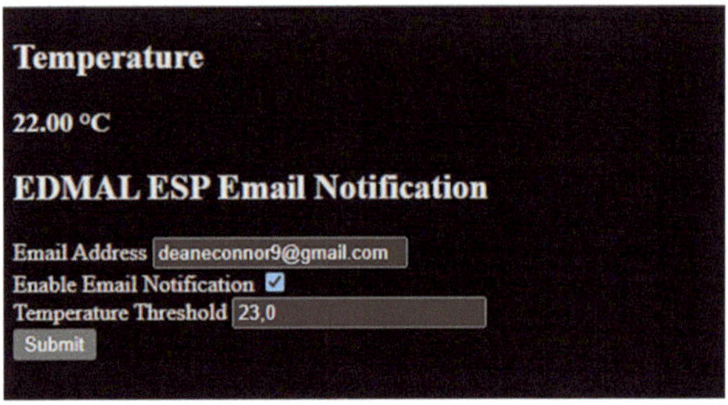

Fig. 12 Web page screenshot for remote access

Remote access testing: Ensured that the system's remote access functionality worked as expected, as seen in Fig. 12 (the designed and access web page). Test remote alerting and control from various locations and devices to confirm seamless connectivity.

Compatibility testing: Assess compatibility with different web browsers and mobile devices to ensure a wide range of users can access and use the system without issues.

5 Results and Discussion

Temperature and humidity data was obtained from simulating various environmental conditions within the enclosed storeroom, including temperature and humidity variations to mimic real world. To ensure the reliability of the developed system data readings, the temperature and humidity sensor had to be calibrated. The precision of the DHT sensor was measured by comparing its output to that of a second sensor displayed from a serial port (Fig. 13).

The simulation of possible environmental conditions within a storeroom considered the system behavior when temperature and humidity threshold values are exceeded. The system should be able to detect and respond to those conditions and trigger an alert. A red RGB LED and a piezoelectric buzzer were able to alert unsafe environmental levels within the storeroom. Figure 14 shows the system output from a serial port displaying the temperature and humidity levels including a timestamp.

Furthermore, the developed system proved ability to recover from a sensor failure. This was achieved by inducing an error on the DHT22 or a network issue, which led to the system signaling failure. The system recovered gracefully as soon as the induced sensor failure was no more. Referring to Fig. 15, the system could

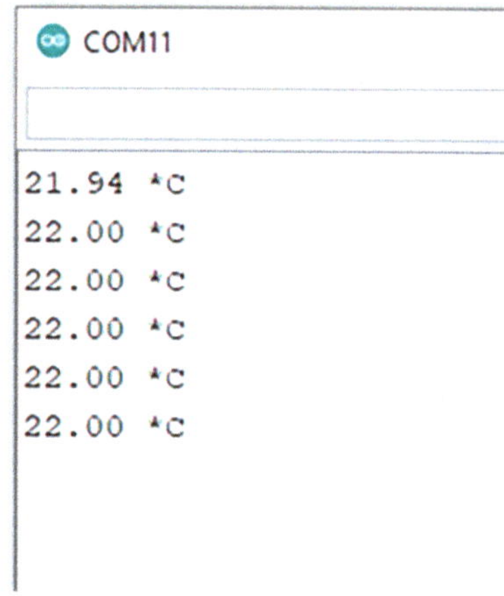

Fig. 13 Second sensor temperature values

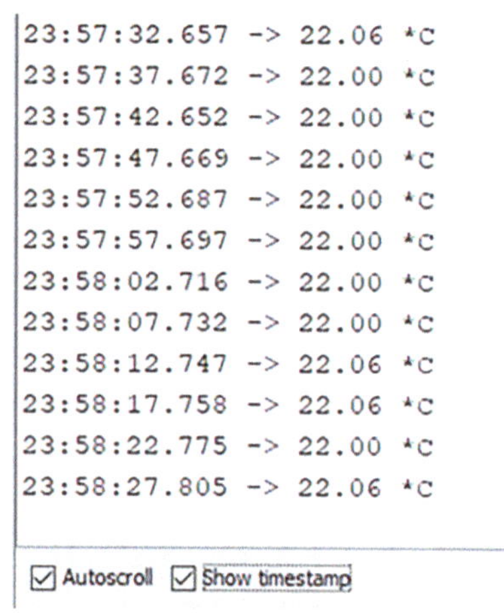

Fig. 14 Data retrieved from the serial port window

recover from failure and continued with operation as seen from the serial port screenshot monitor. The system reboots, finds the IP, and successfully continues taking sensor data samples.

In the testing process, one of the sensors failed. This inspired the researcher to use the DSB1820 calibration sensor not only to perform the duties of calibration and data precision but also to play a role as a secondary sensor that can take over when the primary DHT22 sensor fails. After the system functionality was tested and its results were verified and confirmed reliable, the hardware and software were integrated and 3D printed (Fig. 16).

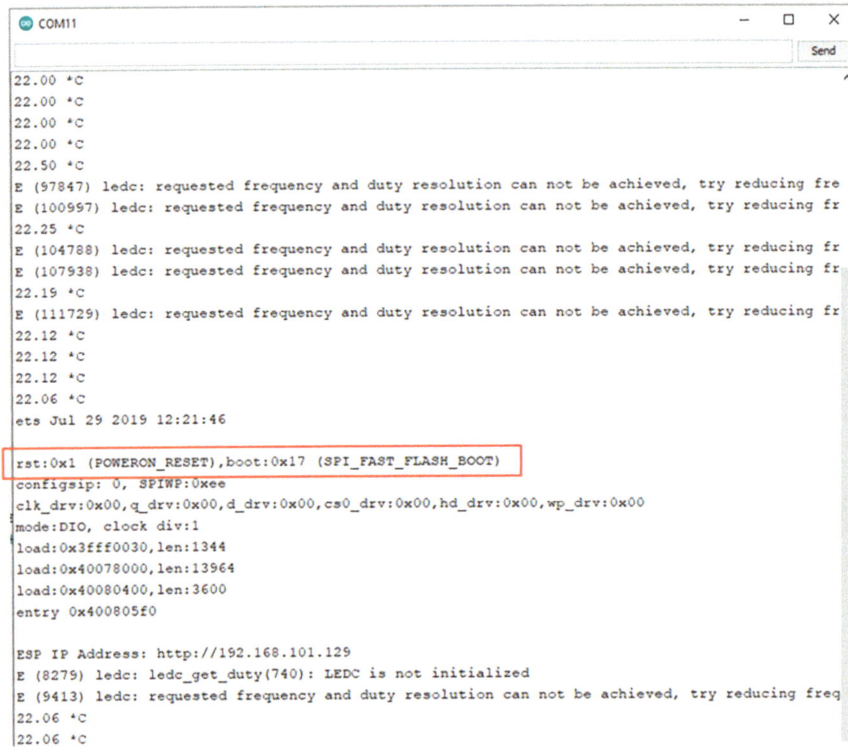

Fig. 15 Reboot notice on serial monitor port window

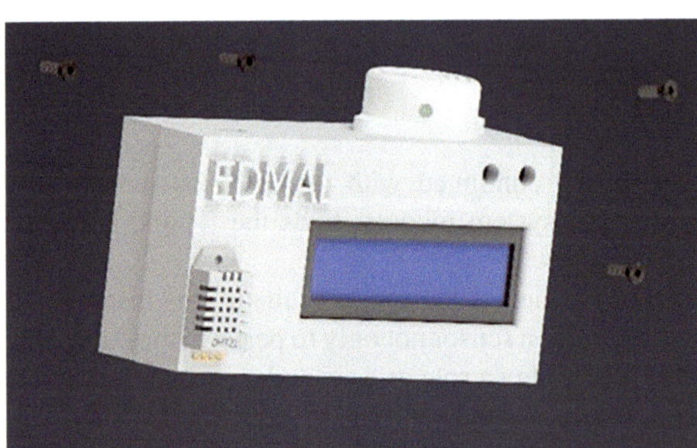

Fig. 16 Final project 3D model illustration

6 Conclusion

Humidity and temperature: The ESP32-based temperature and humidity monitoring system in electronics storeroom was successfully developed through the hardware and software integration phases. Major objectives behind the development of this product included acquiring real-time data, ensuring user-friendliness, providing remote access to the data, and enabling authorized personnel to manage the system from any location with Internet connectivity.

The project outcome was a temperature and humidity data monitoring system enclosed in a 3D printed case that can easily be placed in any storeroom or area for immediate use. The design and development of this ESP32-based monitoring system represent a significant step forward in addressing the critical challenge of safeguarding electronic components from moisture-induced damage. This multifaceted system has demonstrated its capability to deliver comprehensive environmental monitoring and alert functionalities.

References

1. Mazur, B.: IoT for environmental monitoring: reducing costs and enhancing eco-protection. (2024). https://www.ignitec.com/insights/iot-for-environmental-monitoring-reducing-costs-and-enhancing-eco-protection/?print=pdf
2. Katie, B.: Internet of Things (IoT) for environmental monitoring. Int. J. Comput. Eng. **6**(3), 29–42 (2024). https://doi.org/10.47941/ijce.2139
3. Objective, M.: 1 Section 1. Introduction, pp. 1–18.
4. IPC/JEDEC J-STD-033B.1: Handling, Packing, Shipping and Use of Moisture/Reflow Sensitive Surface Mount Devices. Ipc, no. October 2005 (2007)
5. IPC/JEDEC J-STD-020E: Industry Standard Moisture/Reflow Sensitivity Classification for Nonhermetic Surface Mount Devices. JEDEC Solid State Technology Association, no. December (2014)
6. Texas Instruments: Texas Instruments Annual Report, p. 128. (2020). https://investor.ti.com/static-files/05b7598d-4a01-4f45-a63d-058c69a165ad
7. Espressif Systems: ESP32 Series. Esp32, pp. 1–65. (2021). https://www.espressif.com/sites/default/files/documentation/esp32-s2_datasheet_en.pdf
8. Javaid, M., Haleem, A., Rab, S., Pratap Singh, R., Suman, R.: Sensors for daily life: a review. Sens. Int. **2**(July), 100121 (2021). https://doi.org/10.1016/j.sintl.2021.100121
9. Cox, N., Schedtler, D.: Solder reflow basics. In: Reflow Technology Handbook, pp. 1–6. Research International Inc. (2001)
10. Ada, L.: DHT11, DHT22 and AM2302 Sensors (2012). https://learn.adafruit.com/dht/dht-circuitpython-code
11. Mecenas, P., da Rosa Moreira, R.T., Bastos, A.C., Vallinoto, R., Normando, D.: Effects of temperature and humidity on the spread of COVID-19: a systematic review. PLoS One. **15**(9), 1–21 (2020). https://doi.org/10.1371/journal.pone.0238339
12. Yang, J.: Implantable devices. Berkeley Sci. J. **13**(1), 59–60 (2009). https://doi.org/10.5070/bs3131007608
13. Bi, S., Hou, L., Lu, Y.: An integrated wearable strain, temperature and humidity sensor for multifunctional monitoring. Compos. Part A Appl. Sci. Manuf. **149**, 106504 (2021). https://doi.org/10.1016/j.compositesa.2021.106504

14. Srivastava D, Kesarwani A, Dubey S. Measurement of Temperature and Humidity by using Arduino Tool and DHT11. International Research Journal of Engineering and Technology (IRJET) (2018) Dec;5(12):876–8.
15. Zheng, X.X., Li, L.R., Shao, Y.J.: A GSM-based remote temperature and humidity monitoring system for granary. MATEC Web Conf. **44**, 1–5 (2016). https://doi.org/10.1051/matecconf/20164401060
16. Hamed, B., Alami, F.: Adaptive hierarchical fuzzy controller for HVAC systems in low energy buildings. Acad. Platform J. Eng. Sci. **3**(2), 1–7 (2015). https://doi.org/10.5505/apjes.2015.46220
17. Wulandari, E.R.N., Abdullah, H.M., Asri, T.M.: Internet of Things (IoT) Based Temperature and Humidity Monitoring System for Library Rooms. Atlantis Press International BV (2023). https://doi.org/10.2991/978-94-6463-140-1_23
18. Devade, A., Mare, V., Shinde, S., Jori, A., Dilpak, V.: IoT based smart humidity and temperature monitoring system additional by using WiFi module. Int. J. Adv. Res. Innov. Ideas Educ. **8**(3) (2022)
19. Schott, W. An Introduction to Accessing RESTful APIs Using R (2019). https://rpubs.com/plantagenet/481658.
20. Moore, A.: Intro to REST APIs. (2019). https://itcommunity.stanford.edu/sites/default/files/unconference/attachments/2019-11/intro_to_rest_apis.pdf
21. Zlatanov, N.: ARM Architecture and RISC Applications. IEEE Computer Society (2016). https://www.academia.edu/download/44710774/ARM_Architecture_and_RISC_Applications.pdf
22. Spring, O.S.: Today's lecture. Science. **1979**, 1–21 (2006)

Explainable Artificial Intelligence in Dermatology: Overview of Data Scarcity for Skin Cancer Detection in Black Skin

Kavita Behara and Shriya Behara

1 Introduction

Skin cancer is among the most common malignancies worldwide, with millions of new cases identified each year, rendering it a significant public health issue [1]. This cancer includes multiple subtypes, such as basal cell carcinoma, squamous cell carcinoma, and melanoma, with the latter being the most aggressive and potentially lethal if not identified promptly [2]. Early and precise diagnosis in dermatology is essential, as timely management enhances survival chances, reduces the necessity for invasive therapies, and mitigates the financial burden linked to prolonged cancer care [3]. However, a widely reported difficulty continues as individuals from minority populations, especially those with darker skin tones, encounter significant disparities in skin cancer diagnosis and outcome [3].

Despite lower incidences of skin cancer among Skin of Color (SOC) populations, studies indicate that Black individuals and others with darker skin tones (Fitzpatrick skin types IV–VI) face higher mortality rates due to later-stage diagnoses [4]. This delay is attributed to several factors, including limited public awareness, misconceptions about skin cancer risks in SOC populations, and socioeconomic barriers to routine dermatological screenings. Compounding these factors is the pervasive underrepresentation of SOC images in dermatological datasets, which has far-reaching implications for diagnostic technologies that rely on large datasets for accurate and unbiased performance [5–7]. These disparities call for an approach that ensures that SOC populations receive equitable dermatological care,

K. Behara
Mangosuthu University of Technology, Durban, South Africa

S. Behara (✉)
University of Witwatersrand, Johannesburg, South Africa
e-mail: beharak@mut.ac.za

P. Naidoo, M. Sibanda (eds.), *Measurement, Automation, and Control in Artificial Intelligence and Machine Learning*, Signals and Communication Technology, https://doi.org/10.1007/978-3-031-91802-5_12

particularly as medical diagnostics increasingly incorporate artificial intelligence (AI) technologies.

Artificial intelligence, especially deep learning and machine learning models, has demonstrated considerable promise in dermatology, particularly in automating the analysis of medical images to detect and classify skin cancer [8]. AI-driven diagnostic tools, which can process vast quantities of medical data and recognize patterns beyond human capabilities, hold the potential to democratize healthcare by making early diagnostic services accessible even in regions with limited access to dermatologists [9]. However, AI systems in dermatology face an acute challenge: the scarcity of diverse, representative training data that includes SOC populations. AI models trained predominantly on lighter skin types (Fitzpatrick I–III) often struggle to identify malignancies in SOC populations accurately, leading to an increased risk of false negatives and delayed diagnosis [10]. This lack of diversity in training data exacerbates existing healthcare disparities, as individuals from SOC groups are less likely to benefit from the advancements AI offers in dermatological care [11].

One emerging solution to address these disparities involves Explainable AI (XAI), a branch of AI focused on making AI decision-making processes more transparent and interpretable for clinicians and patients [12]. Traditional AI systems, particularly deep learning models, often function as "black boxes," providing accurate predictions without revealing the reasoning behind these decisions [13]. In high-stakes fields like healthcare, this lack of interpretability poses significant risks, as clinicians need to trust and understand the factors influencing AI-driven diagnoses [14]. Explainable AI techniques such as saliency maps, Shapley Additive Explanations (SHAP), and gradient-weighted class activation mapping (Grad-CAM) can elucidate which features are most influential in a model's decision, making it easier for clinicians to assess whether the model is working equitably across different demographic groups, including SOC populations. By providing these insights, XAI can help identify potential biases in AI systems and build trust in AI-assisted dermatology [15].

In addition to XAI, data augmentation techniques, such as synthetic data generation through Generative Adversarial Networks (GANs) and the use of transfer learning, are gaining traction as effective strategies to counteract the data scarcity of SOC images in dermatology [16]. Synthetic data generation, particularly with Deep Convolutional GANs (DCGANs), enables the creation of realistic images representing SOC that can supplement existing datasets. These synthetic images can introduce variability in skin tones, lesion characteristics, and texture, making AI models more resilient and generalizable across diverse populations [17]. Similarly, transfer learning allows for the refinement of pre-trained models on smaller datasets specific to SOC populations, thus reducing the need for large, annotated datasets while improving model performance [18]. These approaches present valuable pathways to reduce bias and increase the inclusivity of AI-driven dermatology tools.

This article provides a comprehensive review of the challenges and implications of data scarcity for SOC populations in skin cancer detection, with a particular focus on the role of XAI in mitigating diagnostic disparities. By examining both existing research and emerging methodologies, this review aims to offer insights into how dermatology can progress toward more accurate and equitable skin cancer detection. It highlights the need for collaborative efforts among AI researchers, dermatologists, and public health stakeholders to create AI models that are inclusive, transparent, and clinically effective for all skin types. Through these efforts, dermatology can advance in both technological sophistication and social equity, ensuring that AI benefits extend to underserved populations who are at a greater risk of diagnostic disparities.

2 The Global Burden of Skin Cancer and Data Disparities

2.1 Global Incidence and Mortality Rates

Skin cancer remains one of the most prevalent cancers worldwide, with significant variations in incidence and mortality rates across different regions. In 2022, an estimated 330,000 new cases of melanoma were diagnosed globally, resulting in nearly 60,000 deaths [19]. Non-melanoma skin cancers (NMSCs), including basal cell carcinoma and squamous cell carcinoma, are even more common. However, precise global statistics are challenging to ascertain due to underreporting and variations in data collection practices. Geographically, countries with predominantly fair-skinned populations and high ultraviolet (UV) exposure, such as Australia and New Zealand, report the highest incidence rates of skin cancer. For instance, Australia had the highest overall rate of skin cancer in 2022, followed by Denmark [20].

Conversely, regions with higher proportions of darker-skinned individuals, such as parts of Africa and Asia, report lower incidence rates. However, these figures may not fully capture the burden of skin cancer in these regions due to underdiagnosis and underreporting.

2.2 Disparities in Diagnosis and Outcomes

Significant disparities exist in skin cancer diagnosis and outcomes among different racial and ethnic groups. Individuals with lighter skin tones are more likely to develop skin cancer, primarily due to lower levels of melanin, which provides some protection against UV radiation. However, people of any race or ethnicity can develop skin cancer, and UV exposure may not be the only risk factor at play [7]. Despite lower incidence rates, individuals with darker skin tones often experience worse outcomes due to later-stage diagnoses. For example, Black and Latino individuals are more likely to be diagnosed with melanoma at advanced stages compared to White individuals, leading to higher mortality rates [6, 7]. Factors contributing to these disparities include lack of awareness, limited access to dermatological care, and underrepresentation in public health campaigns.

2.3 Challenges in Data Collection and Reporting

Accurate data collection and reporting are essential for understanding and addressing skin cancer disparities. However, several challenges hinder this process:

- Underreporting in Low- and Middle-Income Countries (LMICs): Many LMICs lack comprehensive cancer registries, leading to underreporting of skin cancer cases. This underreporting is often due to limited healthcare infrastructure and resources [21].
- Variations in Data Collection Methods: Differences in diagnostic criteria, reporting standards, and healthcare access contribute to inconsistencies in skin cancer data across regions. These variations make it challenging to compare data and develop global strategies [22].
- Underrepresentation of Skin of Color Populations: SOC populations are often underrepresented in dermatological research and datasets, leading to a lack of understanding of skin cancer presentation and outcomes in these groups. This underrepresentation can result in delayed diagnoses and poorer outcomes [10].

2.4 Addressing Data Disparities

To mitigate these disparities, several strategies can be implemented [23]:

- Enhancing data collection infrastructure: Investing in cancer registries and standardized reporting systems, particularly in LMICs, can improve data accuracy and completeness.
- Promoting inclusive research: Encouraging studies that include diverse populations ensures that findings are applicable to all racial and ethnic groups. This inclusivity can lead to a better understanding and management of skin cancer across different demographics.
- Raising awareness and education: Targeted public health campaigns can educate SOC populations about skin cancer risks, symptoms, and the importance of early detection. These campaigns can help reduce disparities in diagnosis and outcomes.

3 Role of AI in Dermatology

Dermatology has a growing history of integrating AI technologies, particularly for skin disease detection and the identification of various skin cancer types. AI tools powered by deep learning algorithms have demonstrated significant value in dermatology due to their ability to automate diagnostic tasks, scale efficiently, and achieve high levels of sensitivity and specificity. By automating the analysis of dermatoscopic images, AI systems effectively reduce the diagnostic workload for dermatologists, allowing for more consistent and timely assessments. These systems can deliver accurate results at scale, making them especially beneficial in settings with limited access to specialized healthcare [24, 25].

Beyond automation, AI's scalability in dermatology offers profound implications for expanding diagnostic capabilities to underserved or resource-limited regions. Through teledermatology and mobile applications, AI can facilitate early diagnosis by providing accessible diagnostic support even where dermatology specialists are scarce. This democratization of healthcare, however, depends on the accuracy and fairness of AI models, which is often compromised by a lack of diverse training data that represents all skin tones, particularly SOC populations [26]. Although AI algorithms have demonstrated diagnostic accuracy, in some cases, exceeds that of experienced dermatologists, their deployment at scale faces limitations due to biases introduced by insufficient training data diversity. These biases can result in inconsistent performance, raising concerns about equity and reliability in AI-driven diagnostics for SOC groups [27].

The potential of AI in dermatology is widely acknowledged, as it not only supports early detection but also promotes cost-efficiency in patient care by reducing the need for invasive procedures and extensive follow-up treatments [27]. However, concerns around fairness persist, particularly for SOC populations, who biased AI models may disproportionately impact. Dermatologists and healthcare professionals recognize the need for ongoing improvements in AI technology to address these disparities, with a focus on ensuring that AI-driven diagnostic tools are equitable, transparent, and capable of supporting diverse patient populations effectively [28].

The ability of an AI model to generalize—to apply learned patterns from training data to new, unseen data—is crucial for its success in clinical applications. In dermatology, this means that AI models must be able to accurately diagnose skin conditions across a wide range of skin tones, ethnicities, and demographic groups. However, the lack of diverse datasets representing SOC populations limits the ability of these models to generalize beyond the demographic groups that dominate their training data [26].

For AI models to effectively generalize darker skin tones, it is essential to include diverse and representative datasets during training. It involves collecting and curating dermatological images that reflect the full spectrum of human skin tones. When models are trained predominantly on white skin, they may fail to recognize critical diagnostic features that differ between light and dark skin. For instance, conditions like hyperpigmentation or melanoma may manifest differently on darker skin, requiring the model to learn distinct patterns for each skin type. Without this diversity in training data, AI models become inflexible and perform poorly in real-world applications involving SOC populations [27].

Moreover, model generalization is not just about training on diverse data but also about ensuring that the algorithms used can handle variability in the data. Robust algorithms that are capable of accounting for differences in skin texture, tone, and lighting conditions are essential for creating AI tools that work effectively across all populations. When models lack diversity in their training datasets, they risk overfitting to the dominant skin types, limiting their utility in diagnosing skin conditions in SOC populations [28].

4 Existing Gaps in AI and Dermatology for SOC Population

AI's transformative potential in dermatology is tempered by significant gaps, particularly in addressing the needs of SOC populations. Despite advancements in AI-driven diagnostics, individuals with darker skin tones remain underserved due to limited representation in dermatology datasets, which can lead to biased models, inaccurate diagnoses, and healthcare disparities. This section examines the critical gaps in AI for dermatology, focusing on data scarcity, diagnostic accuracy, data collection challenges, and label imbalance.

4.1 Data Scarcity and Bias in Dermatology Datasets

One of the significant limitations in applying AI to dermatology is the pervasive issue of data scarcity for SOC populations. Most publicly available dermatology datasets, such as the International Skin Imaging Collaboration (ISIC) archive and HAM10000, predominantly feature lighter skin tones, specifically Fitzpatrick skin types I–III. This lack of representation in training data has significant implications for the diagnostic accuracy and reliability of AI systems when applied to individuals with darker skin tones (Fitzpatrick types IV–VI) [4]. The underrepresentation of SOC images results in models that are trained primarily on lighter skin types, often failing to accurately recognize and classify lesions on darker skin, thus limiting their clinical utility and potentially exacerbating healthcare disparities [7, 11].

4.2 Impact on Diagnostic Accuracy and Healthcare Disparities

The data scarcity issue in SOC populations leads to biases in AI algorithms, which can produce inconsistent and inaccurate diagnostic results for individuals with darker skin. Studies indicate that the characteristics of skin cancers, such as melanoma, often differ in appearance based on skin tone. For example, melanoma on darker skin can present as acral lentiginous melanoma (ALM), typically found in areas like the palms and soles, which are less commonly affected in individuals with lighter skin [23]. AI systems trained predominantly on images of lighter skin types may fail to recognize these specific manifestations, leading to higher rates of false negatives and delayed diagnoses for SOC patients [29, 30]. Such diagnostic inaccuracies are not merely technical challenges but critical health disparities, as they contribute to worse outcomes for individuals with darker skin, who may already face structural barriers to accessing dermatological care.

4.3 Challenges in Data Collection for SOC Populations

Collecting sufficient and diverse dermatology data for SOC populations is challenging due to several factors, including the underreporting of skin conditions in these communities, limited healthcare access, and historical neglect in dermatological research. Many SOC individuals face financial and logistical barriers to receiving timely dermatological care, contributing to underdiagnosis and underrepresentation in dermatology databases. Additionally, even when SOC data is available, it is often limited in quality or lacks detailed annotations, which are essential for training effective AI models [21]. Without standardized guidelines and concerted efforts to include diverse images in dermatology datasets, AI models will continue to fall short of addressing the needs of SOC populations.

4.4 Label Imbalance and Diagnostic Challenges

Beyond image scarcity, label imbalance within datasets further compounds the issue, as malignant lesions on darker skin tones are frequently underrepresented relative to benign conditions. This imbalance can lead to models that overfit to benign presentations, reducing their sensitivity to malignant lesions on SOC skin. For instance, many AI models trained on predominantly benign images of SOC skin may struggle to differentiate between benign conditions like post-inflammatory hyperpigmentation and potentially malignant lesions, increasing the risk of misdiagnosis. Label imbalance also contributes to a lack of generalization, where models perform well on the majority class (lighter skin tones) but poorly on the minority class (SOC populations), ultimately limiting the model's clinical applicability and fairness [17, 22].

4.5 Technical and Ethical Implications of Biased AI Models

The consequences of biased AI in dermatology extend beyond diagnostic inaccuracies; they pose ethical challenges related to fairness and equity in healthcare. Biased AI systems that fail to accurately diagnose SOC populations risk perpetuating historical disparities in dermatology, where individuals with darker skin have faced delayed diagnoses and inadequate care [7]. When AI models systematically underperform SOC populations, they inadvertently reinforce structural inequalities, undermining trust in these technologies among minority groups. Additionally, biases in AI raise questions about accountability in clinical settings, where clinicians and patients alike must trust that AI recommendations are accurate, reliable, and applicable across diverse demographics [31].

4.6 Consequences of Biased Training Data

The consequences of biased training data in dermatology AI are profound, particularly for melanoma detection in darker skin tones. Because AI models trained predominantly on lighter skin tones are not exposed to the distinct characteristics of skin conditions in SOC populations, they are prone to producing false negatives, failing to detect the presence of melanoma or other skin cancers in individuals with darker skin. This bias is especially problematic given that melanoma in SOC populations often presents in atypical locations, such as the palms, soles, and under the nails, and is frequently misdiagnosed as benign.

The use of biased datasets can lead to misdiagnosis and delayed treatment, with profound implications for patient outcomes. For instance, when melanoma is detected at a later stage, the 5-year survival rate drops significantly, particularly for Black individuals, who are often diagnosed at more advanced stages of the disease. It contributes to higher mortality rates in SOC populations, exacerbating existing healthcare disparities. Moreover, a model trained on skewed data may be overfitting to the characteristics of lighter skin tones, making it ill-equipped to handle the variability seen in real-world clinical settings where patients come from diverse backgrounds.

In addition to the clinical risks, biased AI models can also undermine trust in AI-driven healthcare solutions, particularly among minority populations who are already underserved by traditional healthcare systems. Patients who experience misdiagnoses due to biased AI models may be less likely to trust and adopt these technologies, further widening the gap in healthcare access and outcomes between SOC populations and their lighter-skinned counterparts.

5 Explainable AI (XAI) Techniques in Dermatology

The application of XAI in dermatology addresses significant challenges in transparency, trust, and fairness, particularly in high-stakes areas like skin cancer diagnosis. Traditional AI models, especially deep learning-based systems, are often described as "black boxes" due to their opaque decision-making processes, which make it challenging for clinicians to understand or validate how specific diagnoses are reached. This lack of interpretability poses a unique problem in dermatology, where clinical decision-making relies heavily on visible features such as lesion shape, color, texture, and size. XAI techniques, designed to clarify the reasoning behind AI outputs, are essential for facilitating safer, more reliable integration of AI in clinical practice [32].

The interpretability of AI models in dermatology is critical for several reasons. First, it enables dermatologists to scrutinize and validate model predictions against established diagnostic criteria, ensuring that AI recommendations are clinically grounded and appropriate for patient care. Second, transparency fosters trust among clinicians and patients alike, as both groups require assurance that AI systems are not only accurate but also unbiased across diverse patient populations [33, 34]. Additionally, by elucidating how AI models weigh certain features, XAI aids in identifying any racial or demographic biases that may arise due to imbalanced or homogeneous training data. It is crucial for historically underrepresented populations, such as individuals with darker skin, who are often at risk of being underserved by AI tools due to insufficient representation in dermatology datasets [34].

5.1 Key XAI Techniques in Dermatology and Their Applications

XAI techniques, such as SHAP (SHapley Additive exPlanations) or LIME (Local Interpretable Model-agnostic Explanations), are used to interpret and explain the decisions made by AI models.

- LIME: LIME is an adaptable technique that explains complex models by using simpler, interpretable surrogates for individual predictions [35]. In dermatology, LIME helps clarify AI-driven skin lesion classifications by highlighting key visual characteristics—such as asymmetry, border irregularity, or color distribution—that influence a diagnosis. This targeted insight enables dermatologists to assess whether AI focuses on clinically relevant features, enhancing trust and accountability in AI predictions [36].
- SHAP: SHAP, grounded in cooperative game theory, quantifies each feature's contribution to a model's output by assigning "SHAP values," which explain how features like lesion texture or pigmentation influence predictions of malignancy or benignity [37]. SHAP offers both local and global interpretability,

allowing dermatologists to observe individual prediction justifications as well as general patterns across a dataset. In clinical practice, SHAP values can reveal if factors like asymmetry or color variation heavily impact the AI's decision, providing transparency into the model's reasoning. This dual capability of SHAP allows dermatologists to understand broader model tendencies while ensuring each specific case is adequately interpreted according to the unique attributes of the lesion in question [38]. The mathematical equation is as follows:

$$\varnothing_i(f) = \sum_{SCN \setminus \{i\}} \frac{|S|!\,(|N| - |S| - 1)!}{|N|!}\left[f\left(S \cup \{i\}\right) - f(S)\right]$$

- where $f(S)$ is the predicted output of the model for feature set S, N is the set of all features, $\varnothing_i(f)$ is the contribution of feature i to the model's prediction.
- Gradient-weighted Class Activation Mapping (Grad-CAM): Grad-CAM, designed for convolutional neural networks (CNNs), generates heatmaps to highlight image regions most influential in a model's decision, such as irregular borders or dark pigmentation in lesion images [39]. In dermatology, Grad-CAM visually bridges AI analysis and human interpretation, allowing dermatologists to verify if the model's focus aligns with diagnostic features like the ABCDE criteria for melanoma. This validation strengthens the clinical reliability and credibility of AI-assisted diagnoses [40].
- Saliency Maps: Saliency maps highlight the most influential pixels in an image, offering transparency in AI predictions [41]. In skin cancer detection, these maps emphasize critical features like color asymmetry or texture variations, aiding in differentiating malignant from benign lesions [42]. By visually pinpointing areas that affected the AI's classification, saliency maps help clinicians validate model focus, ensuring alignment with diagnostic criteria and reducing the risk of misleading conclusions.

While XAI enhances transparency in AI systems, challenges remain. One key issue is the trade-off between interpretability and accuracy, as simpler models may lose accuracy compared to complex "black box" models. Additionally, reliance on visual explanations from techniques like saliency maps and Grad-CAM can sometimes oversimplify or misrepresent clinical reasoning, leading to misleading interpretations [43]. However, XAI's ability to improve interpretability is invaluable, particularly for Skin of Color (SOC) populations often underrepresented in dermatology datasets. By enhancing transparency, XAI helps detect biases and ensures more equitable performance across diverse skin types, supporting fairer healthcare outcomes.

5.2 Evaluating Effectiveness in Black Skin Detection

One of the biggest challenges in dermatology AI is the limited effectiveness of models trained predominantly on lighter skin tones when applied to Black skin or other SOC populations. XAI techniques offer a way to critically assess how well these models perform across diverse skin tones by highlighting potential biases in their decision-making processes. In evaluating the effectiveness of XAI models for detecting skin cancer in Black skin, several issues arise:

- Inability to fully explain complex AI decisions: While XAI techniques such as saliency maps and SHAP offer insights into AI model behavior, they often fall short of providing complete explanations of how and why a model arrived at its final decision. Complex models, particularly deep learning architectures like CNNs, involve thousands of layers and parameters, making it difficult to unpack the decision-making process fully [32].
- Over-reliance on visual interpretations: Many XAI techniques, particularly saliency maps and CAM, focus heavily on visual explanations. However, these visual cues might not always correlate with the actual underlying features that the model is using to make its decision. It can lead to misleading interpretations of the AI's reasoning, where the saliency map may highlight features that are not clinically significant [32].
- Application to biased datasets: As discussed earlier, one of the most significant challenges in dermatology AI is the lack of diverse data. XAI techniques are only as good as the data on which the models are trained. If the training data is biased—overrepresenting lighter skin tones, for example—then the explanations generated by XAI techniques may also be biased, reflecting the systemic inequalities present in the dataset. It raises concerns about the fairness and equity of XAI models in real-world clinical applications [32].
- Complexity of skin conditions: Skin diseases can vary widely in appearance across different ethnicities and skin tones. The complexity of dermatological conditions poses a challenge for XAI models that may struggle to explain decisions for diseases that manifest differently depending on the patient's skin tone or demographic background [32].
- Interpretability vs. accuracy trade-offs: There is often a trade-off between the accuracy of an AI model and its interpretability. Simpler models, which are easier to explain, may not be as accurate as more complex models. Conversely, highly accurate models, such as deep learning networks, are often less interpretable. Balancing interpretability and performance remains a challenge in deploying XAI in dermatology [32].

6 Advanced Methodologies to Address Data Scarcity

Addressing the lack of diverse data in dermatology, especially for Skin of Color (SOC) populations, is critical for developing AI models that perform equitably across different skin tones. Advanced methodologies like synthetic data generation and transfer learning have emerged as practical solutions to mitigate data scarcity. These techniques enable more inclusive model training, improving generalization and reducing biases in AI-driven dermatology tools.

6.1 *Synthetic Data Generation and Deep Convolutional GANs (DCGANs)*

Synthetic data generation has emerged as a powerful tool in the field of artificial intelligence (AI), particularly for addressing data scarcity issues. In dermatology, where skin types IV–VI (darker skin tones) are severely underrepresented in available datasets, synthetic data generation can help bridge this gap. One of the most effective methods for generating synthetic images is through Generative Adversarial Networks (GANs) and particularly Deep Convolutional GANs (DCGANs) [17].

DCGANs consist of two neural networks, a generator and a discriminator—that work together in a competitive process to produce high-quality synthetic images. The generator creates synthetic skin images, while the discriminator attempts to differentiate between real and synthetic images. Through this iterative process, the generator improves its ability to create images that are increasingly difficult to distinguish from real ones, eventually producing realistic synthetic skin lesion images that can augment existing datasets [17]. The mathematical equation as follows [17]:

$$\min_G \max_D V(G,D) = E_{x\sim P\text{data}(x)}\left[\log(x)\right] + E_{x\sim Pz(z)}\left[\log\left(1 - D\left(G(z)\right)\right)\right]$$

In the context of dermatology AI, DCGANs have proven effective in generating synthetic images that mimic the appearance of skin lesions on darker skin tones, which are often lacking in public datasets. For example, synthetic images of melanoma on Black skin can be created to supplement real-world data, improving the diversity of training datasets. This approach allows AI models to be trained on a more representative set of data, resulting in improved performance and better generalization across different skin types [44].

Advantages of DCGANs

- Diversity: DCGANs can create an infinite number of synthetic images, which allows for an excellent representation of darker skin tones and rare skin conditions.
- Cost-effectiveness: Synthetic data generation is more cost-effective than acquiring real-world dermatological data, which can be time-consuming and expensive.

- Bias mitigation: By augmenting datasets with synthetic images, DCGANs can help mitigate biases in AI models trained predominantly on lighter skin tones, leading to more equitable diagnostic tools.

However, the use of DCGANs is challenging as synthetic images may not perfectly capture the complexity of real-world skin lesions, and there is always a risk that models trained on synthetic data might overfit these images rather than generalizing well to actual clinical data.

6.2 *Transfer Learning and Pre-trained Models*

Transfer learning is another effective methodology for overcoming data scarcity, especially when large, diverse datasets are unavailable. In transfer learning, an AI model is pre-trained on a large, general dataset (often not specific to dermatology) and then fine-tuned on a smaller, more specialized dataset. It allows the model to transfer knowledge from one domain to another, reducing the need for extensive data in the target domain. For example, a model might be pre-trained on ImageNet, a large dataset containing millions of labeled images across various categories. After pre-training, the model can be fine-tuned on a smaller dataset containing dermatological images of darker skin tones. By leveraging the knowledge it has already gained from the larger dataset, the model requires fewer training examples in the specialized domain to achieve high performance [45].

Pre-trained models have been successfully applied in dermatology to improve the detection of skin diseases in SOC populations, even when only limited data is available for these groups. Convolutional Neural Networks (CNNs), such as VGG16 or ResNet, are commonly used in transfer learning due to their ability to capture hierarchical features from images. When fine-tuned on datasets representing darker skin types, these models can achieve better generalization across different skin tones, improving the accuracy of skin cancer detection in Black skin [46].

Advantages of Transfer Learning

- Reduced data requirements: Transfer learning reduces the amount of domain-specific data needed to train a model, making it a practical solution for addressing data scarcity in dermatology.
- Improved performance: Pre-trained models often outperform models trained from scratch, particularly in scenarios where the target dataset is small or imbalanced.
- Faster training: Since the model has already learned general features during pre-training, fine-tuning on a smaller dataset is faster and more efficient.

Despite its advantages, transfer learning has its own set of challenges. For instance, domain mismatch can occur when the source domain (e.g., general image datasets) differs significantly from the target domain (e.g., dermatological images).

Fine-tuning the model to overcome this mismatch can be computationally expensive, especially when working with high-resolution medical images.

6.3 AI-Based Skin Cancer Detection Process

To effectively detect skin cancer in diverse populations, especially SOC, an AI-driven approach is employed in recent literature. This approach follows a precise, structured sequence, as illustrated in Fig. 1. Each step of the process plays a crucial role in ensuring accurate, reliable, and ethically sound skin cancer diagnosis. Below is a detailed narration of the flow:

Step 1: Data Acquisition (Image Collection)

The first step in the process involves the collection of skin images from various sources. This can include clinical dermatology databases or real-world clinical settings. These images form the foundation of the AI model's training process. High-quality, diverse datasets are critical to ensure the model can generalize

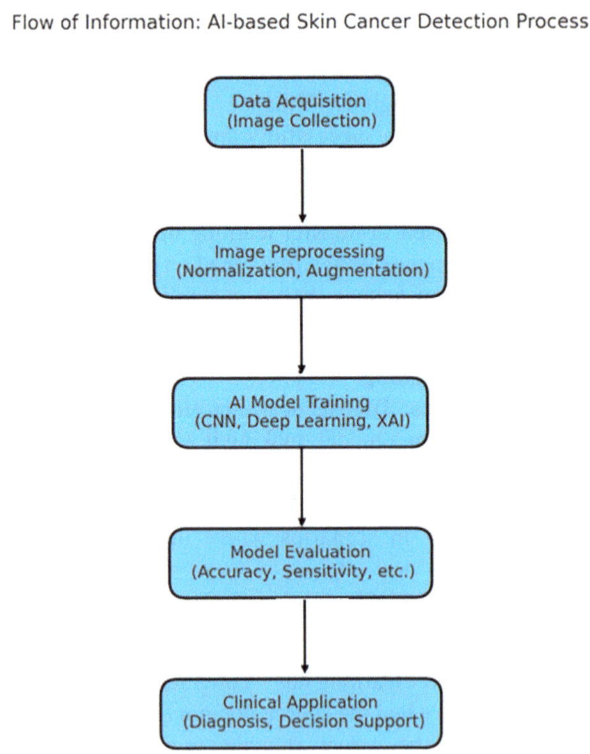

Fig. 1 AI-based detection process for skin cancer

across different skin types, particularly SOC, where data scarcity often hinders accurate detection.

Step 2: Image Preprocessing (Normalization, Augmentation)

Once the images are acquired, they undergo preprocessing to enhance the quality and consistency of the data. Preprocessing steps include normalization (adjusting pixel values to a standard range) and augmentation (creating synthetic variations of images to improve model robustness). This step is crucial for overcoming the challenges posed by dataset imbalances, such as those related to the underrepresentation of SOC populations in the datasets.

Step 3: AI Model Training (CNN, Deep Learning, XAI)

After preprocessing, the next step is training the AI model. Convolutional Neural Networks (CNNs) are typically employed to learn features from the skin images. Deep learning models are also utilized to capture complex patterns in the data. Furthermore, Explainable AI (XAI) techniques are integrated to ensure transparency in the decision-making process, which is particularly important in clinical settings where understanding the model's reasoning is vital for trust and adoption.

Step 4: Model Evaluation (Accuracy, Sensitivity, etc.)

Once the model is trained, its performance is evaluated using key metrics such as accuracy, sensitivity, specificity, and others. This evaluation helps assess the model's effectiveness in distinguishing between malignant and benign lesions, especially across different skin tones. It is critical that the model performs well not just on light skin but on darker skin tones as well, where diagnostic disparities have been reported.

Step 5: Clinical Application (Diagnosis, Decision Support)

The final step in the process is the deployment of the model in a clinical setting. Here, the AI system supports healthcare providers in diagnosing skin cancer. The model can either suggest potential diagnoses or assist dermatologists in decision-making by highlighting suspicious regions in images, which can then be further examined. This step ultimately aims to reduce diagnostic delays and improve outcomes for SOC populations.

With recent advancements and ongoing challenges in the application of AI in dermatology, especially concerning SOC, Table 1 presents a systematic summary of pivotal studies. This summary highlights the evolution of AI applications in dermatology and presents the specific challenges and innovations aimed at enhancing diagnostic accuracy and equity for SOC populations.

Figure 2 presents the graphical analysis of various arguments presented in Table 1.

Table 1 Summary of key research studies of AI in dermatology and skin of color

Author(s)	Year	Study focus	Key findings	Relevance to SOC	Performance metrics	AI models used
Gupta et al.	2016	Skin cancer concerns in people of color	Discussed risk factors and prevention for skin cancer in SOC	Focused on epidemiological insights relevant to SOC		
Tschandl et al.	2020	Human–computer collaboration in skin cancer recognition	Demonstrated effectiveness of combining AI with human expertise in skin cancer ID	Implications for improving AI accuracy in SOC diagnosis	90% accuracy	CNNs
Goon et al.	2021	Skin cancers in skin types IV–VI	Discussed higher risks and later diagnosis in darker skin tones	Highlighted diagnostic disparities in SOC		
Davis et al.	2021	Epidemiology of skin cancer in women of color	Analyzed incidence and clinical manifestations in SOC women	Adds depth to understanding SOC disparities in skin cancer		
Hurtado et al.	2022	Explainable AI in skin lesion diagnosis	Explored methods for making AI in dermatology more explainable and trustworthy	Enhances trust and accuracy in AI diagnostics for SOC		XAI techniques
Dankwa-Mullan et al.	2022	AI, machine learning in cancer care	Addressed the impact of AI and ML on addressing disparities and data diversity	Direct relevance to improving SOC data representation		ML algorithms

(continued)

Table 1 (continued)

Author(s)	Year	Study focus	Key findings	Relevance to SOC	Performance metrics	AI models used
Siegel et al.	2022	Global cancer statistics	Updated statistics on skin cancer incidences worldwide	Contextual background on skin cancer prevalence		
Daneshjou et al.	2022	Disparities in dermatology AI performance	Identified performance disparities in AI systems across diverse clinical image sets	Directly addresses disparities affecting SOC	85% accuracy	CNNs
Brunsgaard et al.	2023	Melanoma in the skin of color	Reviewed epidemiology and clinical presentation of melanoma in SOC	Specific to melanoma diagnosis challenges in SOC	92% sensitivity	CNNs
Brunsgaard et al.	2023	Racial disparities in melanoma outcomes	Analyzed how the late-stage diagnosis affects treatment outcomes in SOC	Highlights critical healthcare gaps affecting SOC populations	88% specificity	CNNs
Jeong et al.	2023	Deep learning in dermatology	Systematically reviewed performance and limitations of AI in dermatology	Broad review and relevance implied for improving SOC diagnosis	87% accuracy	Deep learning models
Luo et al.	2023	AI-assisted dermatology diagnosis	Discussed multimodal AI approaches for improving dermatology diagnosis	Implications for enhancing SOC diagnosis with advanced AI	89% accuracy	Multimodal AI systems
Hossain et al.	2023	Combining pre-trained AI models for skin cancer detection	Showed how ensemble methods improve skin cancer detection using AI	Potential to improve detection accuracy for SOC	93% accuracy	Ensemble models

(continued)

Table 1 (continued)

Author(s)	Year	Study focus	Key findings	Relevance to SOC	Performance metrics	AI models used
Behara et al.	2023	Skin lesion synthesis with DCGAN	Explored synthetic data generation for training AI in dermatology with limited datasets	Provides a method to enhance data representation for SOC	99.38% accuracy	DCGANs
Escalé-Besa et al.	2024	AI in primary care dermatology	Evaluated the use of AI for diagnosing skin diseases in primary care	Demonstrates the potential for AI to support SOC populations	90% accuracy	ML Algorithms
Furriel et al.	2024	AI for clinical skin cancer detection	Reviewed AI's capacity to automate and enhance skin cancer diagnosis in clinical settings	Explores how AI can be adapted to better serve SOC populations	88% accuracy	CNNs
Fliorent et al.	2024	AI advancements in dermatology for SOC	Reviewed AI's challenges in recognizing skin cancer in SOC due to dataset limitations	Specific focus on AI's limitations in SOC		
Kath et al.	2024	Transfer learning in wildlife monitoring	Outlined transfer learning benefits and applications outside healthcare	Analogous insights for applying transfer learning in SOC AI		Transfer learning models
Metta et al.	2024	Interpretable AI in dermatology	Discussed the importance of interpretable AI models in enhancing skin lesion diagnosis	Importance for SOC in ensuring equitable AI diagnostics.		XAI techniques
Goyal et al.	2024	Synthetic data generation in AI	Systematic review of synthetic data techniques in generative AI	Crucial for enhancing SOC dataset representation in AI models		GANs

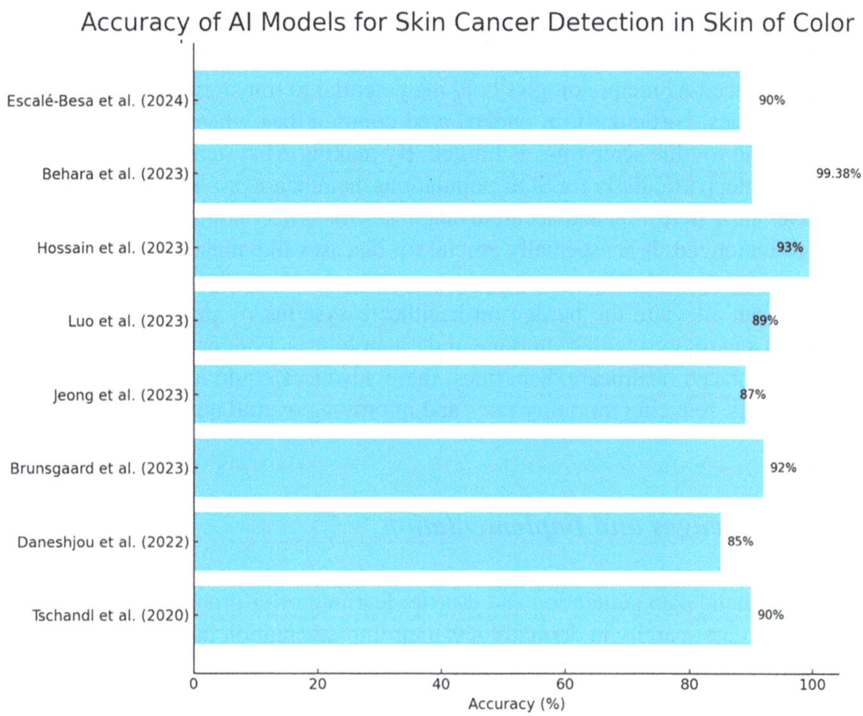

Fig. 2 Accuracy of AI models for skin cancer detection in Skin of Color (SOC) populations from literature

7 Discussion

7.1 Bridging the Gap in Dermatological AI Models

The role of XAI and advanced techniques such as synthetic data generation and transfer learning is critical in addressing the data scarcity challenges faced in dermatological AI. XAI techniques, including LIME, SHAP, and Grad-CAM, offer greater transparency and interpretability of AI models, allowing clinicians to understand the underlying factors driving AI decisions. These techniques are precious when dealing with underrepresented populations, as they help identify biases and areas where the AI may perform inadequately on darker skin tones. In parallel, advanced methods like DCGANs and transfer learning are filling the gap in data availability by generating synthetic images and leveraging pre-trained models to improve generalization across diverse skin types. The combined application of these technologies is enabling AI models to become more inclusive, equitable, and effective in diagnosing skin conditions for SOC populations, thereby contributing to fairer healthcare outcomes.

7.2 Potential Impact on Clinical Practice

These advanced AI methodologies hold the potential to transform real-world medical diagnostics, particularly in underserved communities where access to dermatologists and routine screenings is limited. By making AI systems more robust and generalizable, particularly for SOC populations, healthcare providers can rely on AI to deliver early detection and accurate diagnoses for skin conditions that may otherwise go unnoticed. It is especially crucial for diseases like melanoma, where early detection significantly improves patient outcomes. Additionally, AI-driven diagnostic tools can alleviate the burden on healthcare systems by providing automated diagnostics in regions with a shortage of dermatologists. For communities that have historically faced healthcare disparities, these advances could lead to more timely interventions, reducing mortality rates and improving overall quality of care.

7.3 Challenges and Implementation

While synthetic data generation and transfer learning offer promising solutions for addressing data scarcity in dermatology, their implementation in real-world clinical settings is fraught with challenges.

- Ethical concerns: The use of synthetic data raises ethical questions about the authenticity and validity of the generated images. Clinicians may be hesitant to trust AI models trained on synthetic data, especially if the images do not perfectly replicate the nuances of real skin lesions. Additionally, synthetic data must be carefully labelled and validated to ensure it accurately represents the target population, including darker skin tones.
- Computational complexity: Both DCGANs and transfer learning require significant computational resources. Training a DCGAN to generate high-quality synthetic images or fine-tuning a pre-trained model can be resource-intensive, requiring access to high-performance computing infrastructure. It can be a barrier to adoption, particularly in low-resource settings where such infrastructure may not be readily available.
- Bias and overfitting: While these techniques can help address data imbalance, they are not immune to the risk of overfitting or perpetuating biases present in the original training data. For example, suppose the synthetic images generated by DCGANs are not diverse enough or fail to capture the full range of variability in skin lesions. In that case, the AI model may still struggle to generalize to real-world cases. Similarly, transfer learning models may retain biases from the original pre-training dataset, limiting their effectiveness in diagnosing skin conditions in underrepresented populations.
- Data privacy and security: Federated learning offers a potential solution to data scarcity by allowing multiple institutions to collaborate on AI model development without sharing sensitive patient data. However, the implementation of

federated learning requires robust data privacy and security protocols to ensure that patient data is protected during model training. It is essential in healthcare, where breaches of patient confidentiality can have profound legal and ethical implications.

- Resource limitations: Healthcare institutions in LMICs may face resource constraints that hinder their ability to implement advanced AI techniques. Lack of access to high-quality imaging equipment, annotated datasets, and computational power can limit the potential for AI-driven dermatology tools in these regions, perpetuating existing healthcare disparities.

7.4 Limitations

Despite the promising nature of these advanced techniques, several limitations remain. Synthetic data generation, while helpful, can struggle to replicate the complexity and variability of real-world skin conditions fully. It can result in overfitting to synthetic data and limit the AI's performance in actual clinical settings. Similarly, transfer learning depends heavily on the quality and relevance of the pre-trained models, which may carry biases from their source datasets. Furthermore, the implementation of techniques like federated learning poses challenges related to data privacy, computational resources, and infrastructure, particularly in low-resource settings. To overcome these limitations, future research must focus on developing more diverse datasets, refining synthetic data techniques, and improving collaborative models like federated learning. Additionally, ethical concerns related to algorithmic fairness and patient safety must be addressed to ensure these technologies benefit all populations.

8 Conclusion and Future Scope

This article has focused on the critical challenges posed by data scarcity in dermatological AI, particularly with regard to SOC populations. The underrepresentation of darker skin tones in training datasets has resulted in AI models that demonstrate suboptimal performance when diagnosing conditions in SOC populations, thereby contributing to misdiagnoses and exacerbating healthcare disparities. However, the study has identified several promising solutions, such as the application of XAI techniques, synthetic data generation, and transfer learning, which can potentially address these limitations and enhance AI performance across all skin types. The development of inclusive AI models for dermatology is not only an ethical imperative but also a practical necessity to ensure equitable healthcare outcomes. AI systems must be designed to accurately detect and diagnose skin conditions across a diverse range of skin tones. It is essential for ensuring that underserved populations receive the same level of diagnostic accuracy and care as those in better-represented groups. In this context, the dermatology community, in collaboration with AI researchers and healthcare policymakers, must prioritize the development of diverse datasets and implement bias mitigation strategies to foster the creation of equitable AI solutions.

Key recommendations for future research and practice include:

- Expansion of diverse dermatological datasets: The availability of diverse datasets, particularly for SOC populations, should be prioritized. Collaborative efforts between hospitals, research institutions, and AI developers are crucial to ensure that data-sharing initiatives, such as federated learning, become more widespread while maintaining privacy standards.
- Establishment of regulatory frameworks: Policymakers must establish comprehensive regulatory frameworks that ensure AI models meet fairness and bias standards. These models should undergo rigorous audits for bias before being deployed in clinical settings.
- Clinician engagement with XAI: It is essential to encourage clinicians to engage with XAI tools. It will allow healthcare providers better to understand the strengths and limitations of AI diagnostics, leading to more informed and effective clinical decision-making.

In conclusion, by following these recommendations, it is possible to advance the development of AI-driven dermatology that is inclusive, equitable, and clinically valuable for all patients, ensuring improved healthcare outcomes and a reduction in existing disparities.

References

1. Siegel, R.L., Miller, K.D., Fuchs, H.E., Jemal, A.: Cancer statistics, 2022. CA Cancer J. Clin. **72**, 7–33 (2022). https://doi.org/10.3322/caac.21708
2. Roky, A.H., Islam, M.M., Ahasan, A.M.F., Mostaq, M.S., Mahmud, M.Z., Amin, M.N., Mahmud, M.A.: Overview of skin cancer types and prevalence rates across continents. Cancer Pathog. Ther. (2024). https://doi.org/10.1016/j.cpt.2024.08.002
3. Omotoso, O., Teibo, J.O., Atiba, F.A., Oladimeji, T., Paimo, O.K., Ataya, F.S., Batiha, G.E.-S., Alexiou, A.: Addressing cancer care inequities in sub-Saharan Africa: current challenges and proposed solutions. Int. J. Equity Health. **22**, 189 (2023). https://doi.org/10.1186/s12939-023-01962-y
4. Goon, P., Banfield, C., Bello, O., Levell, N.J.: Skin cancers in skin types IV–VI: does the Fitzpatrick scale give a false sense of security? Skin Health Dis. **1** (2021). https://doi.org/10.1002/ski2.40
5. Davis, D.S., Robinson, C., Callender, V.D.: Skin cancer in women of color: epidemiology, pathogenesis and clinical manifestations. Int. J. Womens Dermatol. **7**, 127–134 (2021). https://doi.org/10.1016/j.ijwd.2021.01.017
6. Fajuyigbe, D., Young, A.R.: The impact of skin colour on human photobiological responses. Pigment Cell Melanoma Res. **29**, 607–618 (2016). https://doi.org/10.1111/pcmr.12511
7. Lin, R.R., Lee, J., Maderal, A.D., Elman, S.A.: Rural Health Disparities in Skin Cancer Amplified Among Skin of Color (2024)

8. Jeong, H.K., Park, C., Henao, R., Kheterpal, M.: Deep learning in dermatology: a systematic review of current approaches, outcomes, and limitations. JID Innov. **3**, 100150 (2023). https://doi.org/10.1016/j.xjidi.2022.100150
9. Luo, N., Zhong, X., Su, L., Cheng, Z., Ma, W., Hao, P.: Artificial intelligence-assisted dermatology diagnosis: from unimodal to multimodal. Comput. Biol. Med. **165**, 107413 (2023). https://doi.org/10.1016/j.compbiomed.2023.107413
10. Fliorent, R., Fardman, B., Podwojniak, A., Javaid, K., Tan, I.J., Ghani, H., Truong, T.M., Rao, B., Heath, C.: Artificial intelligence in dermatology: advancements and challenges in the skin of color. Int. J. Dermatol. **63**, 455–461 (2024). https://doi.org/10.1111/ijd.17076
11. Daneshjou, R., Vodrahalli, K., Novoa, R.A., Jenkins, M., Liang, W., Rotemberg, V., Ko, J., Swetter, S.M., Bailey, E.E., Gevaert, O., Mukherjee, P., Phung, M., Yekrang, K., Fong, B., Sahasrabudhe, R., Allerup, J.A.C., Okata-Karigane, U., Zou, J., Chiou, A.S.: Disparities in dermatology AI performance on a diverse, curated clinical image set. Sci. Adv. **8** (2022). https://doi.org/10.1126/sciadv.abq6147
12. Sadeghi, Z., Alizadehsani, R., Akif CIFCI, M., Kausar, S., Rehman, R., Mahanta, P., Bora, P.K., Almasri, A., Alkhawaldeh, R.S., Hussain, S., Alatas, B., Shoeibi, A., Moosaei, H., Hladík, M., Nahavandi, S., Pardalos, P.M.: A review of Explainable Artificial Intelligence in healthcare. Comput. Electr. Eng. **118**, 109370 (2024). https://doi.org/10.1016/j.compeleceng.2024.109370
13. Praveenraj, D.D.W., Victor, M., Vennila, C., Hussein Alawadi, A., Diyora, P., Vasudevan, N., Avudaiappan, T.: Exploring explainable artificial intelligence for transparent decision making. E3S Web Conf. **399**, 04030 (2023). https://doi.org/10.1051/e3sconf/202339904030
14. Saeed, W., Omlin, C.: Explainable AI (XAI): a systematic meta-survey of current challenges and future opportunities. Knowl.-Based Syst. **263**, 110273 (2023). https://doi.org/10.1016/j.knosys.2023.110273
15. Abgrall, G., Holder, A.L., Chelly Dagdia, Z., Zeitouni, K., Monnet, X.: Should AI models be explainable to clinicians? Crit. Care. **28**, 301 (2024). https://doi.org/10.1186/s13054-024-05005-y
16. Goyal, M., Mahmoud, Q.H.: A systematic review of synthetic data generation techniques using generative AI. Electronics. **13**, 3509 (2024). https://doi.org/10.3390/electronics13173509
17. Behara, K., Bhero, E., Agee, J.T.: Skin lesion synthesis and classification using an improved DCGAN classifier. Diagnostics. **13**, 2635 (2023). https://doi.org/10.3390/diagnostics13162635
18. Kath, H., Serafini, P.P., Campos, I.B., Gouvêa, T.S., Sonntag, D.: Leveraging transfer learning and active learning for data annotation in passive acoustic monitoring of wildlife. Ecol. Inform. **82**, 102710 (2024). https://doi.org/10.1016/j.ecoinf.2024.102710
19. IARC—International Agency for Research On Cancer. https://www.iarc.who.int/
20. Cancer Prevention Organisation: World Cancer Research Fund International—WCRF International. https://www.wcrf.org/
21. Behara, K., Bhero, E., Agee, J.T., Gonela, V.: Artificial intelligence in medical diagnostics: a review from a South African context. Sci. Afr. **17**, e01360 (2022). https://doi.org/10.1016/j.sciaf.2022.e01360
22. Tschandl, P., Rinner, C., Apalla, Z., Argenziano, G., Codella, N., Halpern, A., Janda, M., Lallas, A., Longo, C., Malvehy, J., Paoli, J., Puig, S., Rosendahl, C., Soyer, H.P., Zalaudek, I., Kittler, H.: Human-computer collaboration for skin cancer recognition. Nat. Med. **26**, 1229–1234 (2020). https://doi.org/10.1038/s41591-020-0942-0
23. Gupta, A.K., Bharadwaj, M., Mehrotra, R.: Skin cancer concerns in people of color: risk factors and prevention. Asian Pacific J. Cancer Prev. **17**, 5257–5264 (2016). https://doi.org/10.22034/APJCP.2016.17.12.5257
24. Furriel, B.C.R.S., Oliveira, B.D., Prôa, R., Paiva, J.Q., Loureiro, R.M., Calixto, W.P., Reis, M.R.C., Giavina-Bianchi, M.: Artificial intelligence for skin cancer detection and classification for the clinical environment: a systematic review. Front. Med. **10**, 1305954 (2024). https://doi.org/10.3389/fmed.2023.1305954

25. Brancaccio, G., Balato, A., Malvehy, J., Puig, S., Argenziano, G., Kittler, H.: Artificial intelligence in skin cancer diagnosis: a reality check. J. Invest. Dermatol. **144**, 492–499 (2024). https://doi.org/10.1016/j.jid.2023.10.004
26. Escalé-Besa, A., Vidal-Alaball, J., Miró Catalina, Q., Gracia, V.H.G., Marin-Gomez, F.X., Fuster-Casanovas, A.: The use of artificial intelligence for skin disease diagnosis in primary care settings: a systematic review. Healthcare. **12**, 1192 (2024). https://doi.org/10.3390/healthcare12121192
27. Haykal, D.: The suitability of AI in dermatology for enhanced skin care. J. Aesthetic Nurs. **13** (2024)
28. Dankwa-Mullan, I., Weeraratne, D.: Artificial intelligence and machine learning technologies in cancer care: addressing disparities, bias, and data diversity. Cancer Discov. **12**, 1423–1427 (2022). https://doi.org/10.1158/2159-8290.CD-22-0373
29. Brunsgaard, E.K., Wu, Y.P., Grossman, D.: Melanoma in the skin of color: Part I. Epidemiology and clinical presentation. J. Am. Acad. Dermatol. **89**, 445–456 (2023). https://doi.org/10.1016/j.jaad.2022.04.056
30. Brunsgaard, E.K., Jensen, J., Grossman, D.: Melanoma in skin of color: Part II. Racial disparities, role of UV, and interventions for earlier detection. J. Am. Acad. Dermatol. **89**, 459–468 (2023). https://doi.org/10.1016/j.jaad.2022.04.057
31. Ueda, D., Kakinuma, T., Fujita, S., Kamagata, K., Fushimi, Y., Ito, R., Matsui, Y., Nozaki, T., Nakaura, T., Fujima, N., Tatsugami, F., Yanagawa, M., Hirata, K., Yamada, A., Tsuboyama, T., Kawamura, M., Fujioka, T., Naganawa, S.: Fairness of artificial intelligence in healthcare: review and recommendations. Jpn. J. Radiol. **42**, 3–15 (2024). https://doi.org/10.1007/s11604-023-01474-3
32. Hurtado, S., Nematzadeh, H., García-Nieto, J., Berciano-Guerrero, M.Á., Navas-Delgado, I.: On the use of explainable artificial intelligence for the differential diagnosis of pigmented skin lesions. Lect. Notes Comput. Sci. **13346 LNBI**, 319–329 (2022). https://doi.org/10.1007/978-3-031-07704-3_26
33. Metta, C., Beretta, A., Guidotti, R., Yin, Y., Gallinari, P., Rinzivillo, S., Giannotti, F.: Advancing dermatological diagnostics: interpretable AI for enhanced skin lesion classification. Diagnostics. **14**, 753 (2024). https://doi.org/10.3390/diagnostics14070753
34. Ghassemi, M., Oakden-Rayner, L., Beam, A.L.: The false hope of current approaches to explainable artificial intelligence in health care. Lancet Digit. Health. **3**, e745–e750 (2021). https://doi.org/10.1016/S2589-7500(21)00208-9
35. Ribeiro, M.T., Singh, S., Guestrin, C.: Why should I trust you? In: Proceedings of the 22nd ACM SIGKDD International Conference on Knowledge Discovery and Data Mining, pp. 1135–1144. ACM, New York, NY (2016)
36. Attallah, O.: Skin-CAD: explainable deep learning classification of skin cancer from dermoscopic images by feature selection of dual high-level CNNs features and transfer learning. Comput. Biol. Med. **178**, 108798 (2024). https://doi.org/10.1016/j.compbiomed.2024.108798
37. Lundberg, S.M., Lee, S.I.: A unified approach to interpreting model predictions. Adv. Neural Inf. Process. Syst. **2017**, 4766–4775 (2017)
38. Khater, T., Ansari, S., Mahmoud, S., Hussain, A., Tawfik, H.: Skin cancer classification using explainable artificial intelligence on pre-extracted image features. Intell. Syst. Appl. **20**, 200275 (2023). https://doi.org/10.1016/j.iswa.2023.200275
39. Selvaraju, R.R., Cogswell, M., Das, A., Vedantam, R., Parikh, D., Batra, D.: Grad-CAM: visual explanations from deep networks via gradient-based localization. In: 2017 IEEE International Conference on Computer Vision (ICCV), pp. 618–626. IEEE (2017)
40. Matas, I., Serrano, C., Silva-Clavería, F., Serrano, A., Toledo-Pastrana, T., Acha, B.: AI-Driven Skin Cancer Diagnosis: Grad-CAM and Expert Annotations for Enhanced Interpretability. (2024)
41. de Brecht, M., Saiki, J.: A neural network implementation of a saliency map model. Neural Netw. **19**, 1467–1474 (2006). https://doi.org/10.1016/j.neunet.2005.12.004

42. Toteja, R., Gupta, D., Gautam, V., Vishwakarma, D.K.: Skin disease detection using saliency maps and segmentation techniques. In: Presented at the Communications in Computer and Information Science (2023)
43. Sundararajan, M., Taly, A., Yan, Q.: (Integrated Gradient) Axiomatic attribution for deep networks. In: 34th Int. Conf. Mach. Learn, vol. 7, pp. 5109–5118 (2017)
44. Nirmala, V., Premaladha, J.: Synthetic medical image augmentation: a GAN-based approach for melanoma skin lesion classification with deep learning. In: Deep Learning in Personalized Healthcare and Decision Support, pp. 69–80. Elsevier (2023)
45. Dahou, A., Aseeri, A.O., Mabrouk, A., Ibrahim, R.A., Al-Betar, M.A., Elaziz, M.A.: Optimal skin cancer detection model using transfer learning and dynamic-opposite hunger games search. Diagnostics. **13**, 1579 (2023). https://doi.org/10.3390/diagnostics13091579
46. Hossain, M.M., Hossain, M.M., Arefin, M.B., Akhtar, F., Blake, J.: Combining state-of-the-art pre-trained deep learning models: a noble approach for skin cancer detection using max voting ensemble. Diagnostics. **14**, 89 (2023). https://doi.org/10.3390/diagnostics14010089

Quantum-Enhanced Neural Network for Forecasting Kenyan Economic Growth

Saif Kinyori, Edna C. Too, David G. Mwathi, and Juster N. Mungiria

1 Introduction

Technology keeps on changing at enormous speed, and the generation of complex sets of data is on the rise [10]. In order to find hidden patterns, correlations, and other insights, data analytics investigates these complex or massive amounts of data [20]. Finding knowledge from unstructured data gathered from the Internet is difficult since this material is intended for human consumption rather than machine consumption. Additionally, exploiting these incredibly complex datasets requires a significant amount of training, testing, and validation effort. Data avenues have the potential to provide value over time by enhancing things like smart tourist destinations [6]. Accordingly, data gathering from user-generated material, together with its aggregation, interconnection, analysis, and integration, real-time synchronization, and intelligent use of data, can be used to create a smart experience [17].

Furthermore, data analysis offers substantial benefits within the public and governmental domains. Governments that opt to share their extensive datasets foster the exchange of information, enhance transparency, and cultivate trust among their constituents [21]. Leveraging data provided by government agencies, encompassing aspects like crops, weather, and geography, holds the potential to enhance existing approaches in agriculture and industry. Similarly, the examination of hospital admission data in conjunction with demographic and census information has the potential to enhance healthcare effectiveness in underserved regions [11].

S. Kinyori (✉) · E. C. Too · D. G. Mwathi · J. N. Mungiria
Chuka University, Chuka, Kenya
e-mail: skinyori@chuka.ac.ke; echebt@chuka.ac.ke; dgmwathi@chuka.ac.ke; jmungiria@chuka.ac.ke

P. Naidoo, M. Sibanda (eds.), *Measurement, Automation, and Control in Artificial Intelligence and Machine Learning*, Signals and Communication Technology, https://doi.org/10.1007/978-3-031-91802-5_13

Researchers are using more complex techniques to comprehend and make sense of the exponentially growing amount of data [24]. In this analysis, artificial intelligence, a subfield of computer science that deals with giving machines intelligence, has shown considerable promise in recent years [9]. Data analysis has relied heavily on machine learning methods [17]. Artificial neural networks have made considerable advances in big and complex data processing over the years, and this has had a substantial impact on the field of deep learning. In data analytics with artificial neural networks, huge or complex volumes of data are involved in research and researchers have to perform cross-validation to archive the best optimal results. However, this process takes a lot of time running on classical machines. Cross-validation is the process where a repetition of training datasets is performed a number of times and the accuracy results are weighted [19].

Field-programmable gate arrays (FPGAs), application-specific integrated circuits (ASICs), and graphics processing units (GPUs) that are specifically designed for machine learning are special machine learning chips that are employed in data analytics. Like all technologies, there are advantages and trade-offs to every solution. Overall, researchers have to choose the best technology based on their specific end application as these chips perform better in specific environments and applications [19]. The amount of data that need to be stored and processed in machine learning chips is much larger than that used in common applications before [23]. This creates a huge demand and is reflected not only in the demand of machine learning applications, but also in the new computing paradigm required by machine learning, especially deep learning [18]. The demands require more efficient hardware to process intelligent computing and analysis, and we also encounter some bottlenecks in the current technical framework, especially the von Neumann bottlenecks, CMOS process, and device bottlenecks. Therefore, most of the hardware architecture innovations for machine learning, especially for accelerating neural network processing, are struggling with this problem. This presents an opportunity to explore the quantum realm.

Quantum computers harness certain enigmatic properties of quantum mechanics [2], a scientific realm delving into the behavior of matter and light on the atomic and subatomic level, to bring about significant advancements in processing capability. Quantum machines are anticipated to surpass even the mightiest classical supercomputers [4]. Traditional computers rely on bits, which are sequences of electrical or optical signals representing 1s or 0s. In contrast, quantum computers employ qubits, often subatomic particles like electrons or photons [25]. A grouped ensemble of qubits can process information far more expeditiously than an equivalent number of binary bits [5]. These properties, specifically entanglement and superposition, distinguish qubits [25].

Quantum computers utilize quantum bits or qubits to represent and process information, guided by the principles of quantum mechanics [14]. These computers have the ability to operate within an immensely expanded computational domain, with resource requirements growing at a polynomial rather than exponential rate, as dictated by specific quantum phenomena [3]. Algorithms that can be effectively executed on quantum computers hold the promise of achieving exponential speedup compared to the most advanced classical methods currently in use.

GDP is a fundamental indicator used to gauge the overall economic performance of a country or region [1]. In sub-Saharan Africa, GDP plays a vital role in assessing economic growth, income distribution, and living standards [8]. Countries in this region face unique challenges including poverty, political instability, and inadequate infrastructure, which can significantly impact GDP growth rates [7]. Studies have examined the factors influencing GDP in sub-Saharan Africa and the impact of foreign direct investment (FDI) and trade openness on GDP growth in selected countries in sub-Saharan Africa [16]. The results show that FDI and trade openness have positive effects on GDP growth, highlighting the importance of external economic factors in the region.

Economic forecasting is a critical process that helps policymakers, investors, and businesses make informed decisions [13]. Accurate economic forecasts can lead to better resource allocation, increased investments, and overall economic stability [12]. However, forecasting in sub-Saharan Africa poses unique challenges because of the region's economic and political complexities [15]. Studies have explored various economic forecasting models to enhance the prediction accuracy. Machine learning techniques have been utilized to forecast GDP growth in selected countries in sub-Saharan Africa. The studies have also demonstrated the potential of machine learning algorithms for improving forecasting accuracy in the region.

2 Methodology

2.1 Data Sampling

The research study used the Kenyan economy data from World Bank datasets. The World Bank database provides a comprehensive collection of economic data [22]. The data comprised areas such as social media market trends, web traffic search, financial transactions, Kenya poverty metrics, Kenyan oil prices, agricultural and household items pricing, Kenyan supermarkets data, economic impact of Covid-19, and other related economic factors (Table 1).

2.2 Data Preprocessing

The preprocessing of data involved cleaning, transforming, and organizing data into a structured format. The dataset had inconsistencies, errors, and missing values that needed to be addressed before it was used for training. The preprocessing step included tasks of data cleaning, data transformation, normalization, and outlier

Table 1 Sampled dataset of Kenyan economic indicators

Country name	Country code	Indicator name	Indicator code	1960	1961	1962
Kenya	KEN	Merchandise exports to low- and middle-income	TX.VAL. MRCH. WR.ZS	8.113 035552	9.694 323144	9.5316 35168
Kenya	KEN	Merchandise exports to low- and middle-income economies in South Asia	TX.VAL. MRCH. R5.ZS	4.375 569736	4.366 812227	3.451 109285
Kenya	KEN	Merchandise exports to low- and middle-income economies in East Asia and Pacific	TX.VAL. MRCH. R1.ZS	0.9115 77028	0.2620 08734	0.5751 84881
Kenya	KEN	Merchandise exports to economies in the Arab World	TX.VAL. MRCH. AL.ZS	3.1905 19599	3.231 441048	3.3689 40016
Kenya	KEN	Merchandise imports from low- and middle-income economies in Sub-Saharan Africa	TM.VAL. MRCH. R6.ZS	4.7676 28205	5.630 908344	4.2738 42501
Kenya	KEN	Merchandise imports (current US$)	TM.VAL. MRCH. CD.WT	196000 000 0000	19300 0000	19500 0000

detection. Preprocessing the data ensured that the data is consistent, complete, and ready to be used for training of the proposed model.

2.3 System Modeling

To establish entanglement between qubits and generate unique circuit pairs for model processing, a combination of a CNOT gate and a Hadamard gate was employed. The design incorporated a distinct architectural configuration where the neuron count within the encoder layer progressively diminished across successive layers. Conversely, as the model progressed to subsequent layers, the neuron count within the decoder layer experienced an increment. This structured approach encompassed three tiers for both the encoder and decoder components. The encoder contained 64, 32, and 1 units, respectively, in each level, and the decoder contained 1, 32, and 64 units, respectively, in each level. This provided a symmetric approach. The data was also scaled between 0 and 1 before feeding it into the autoencoder using minimum maximum scaler as the process employed different activation functions in the output layer which outputs values between 0 and 1.

The modeling procedure encompassed inputting historical data into the quantum-enhanced neural network, refining the weights, and fine-tuning parameters to minimize the disparity between predicted and actual outputs. This process incorporated the utilization of backpropagation and gradient descent methods. By training the model, the algorithm acquired the ability to discern patterns within the historical data, facilitating precise forecasting.

The model's effectiveness was assessed through metrics including mean absolute error, mean squared error, and root mean squared error. Additionally, cross-validation was employed to validate the model's efficiency, and the time taken for each iteration was documented.

3 Results and Discussion

The model summary in Fig. 1 shows the layers, output shape, and number of parameters of the quantum-enhanced neural network model (Fig. 2).

3.1 Model Design

See Figs. 3 and 4.

3.2 Model Summary

The validation loss in Fig. 2 shows how well the model fits new data, while the training loss shows how well it matches training data. An Adam optimizer was used to produce the training and validation loss, as illustrated in Fig. 3 as well.

Epochs	ReLU/SDG Loss	ReLU/SDG Time(s)	Sigmoid/SDG Loss	Sigmoid/ SDG Time(s)	Tanh/SDG Loss	Tanh/SDG Time(s)
0	0.25	0.5	0.56	0.5	0.55	0.7
250	0.25	1.5	0.5	0.9	0.54	1.3
500	0.2	1.6	0.5	1.3	0.5	1.5
750	0.15	1.9	0.43	1.5	0.45	1.9
1000	0.15	2.3	0.4	1.7	0.44	2.4
1250	0.1	2.6	0.36	1.9	0.35	2.9
1500	0.1	2.9	0.35	2.4	0.25	3.2
1750	0.1	2.9	0.3	2.5	0.2	3.4
2000	0	3.4	0.25	2.7	0.15	3.7

Fig. 1 Stochastic gradient descent optimizer vs. ReLU, sigmoid and Tan*h* activation functions

Epochs	ReLU/Adam Loss	ReLU/Adam Time(s)	Sigmoid/Adam Loss	Sigmoid/Adam Time(s)	Tanh/Adam Loss	Tanh/Adam Time(s)
0	0.25	0	1.2	0.7	1.3	0.9
250	0.25	0.5	1	0.9	1.2	1.2
500	0.2	0.5	0.7	1.2	0.95	1.4
750	0.15	1	0.6	1.4	0.8	1.7
1000	0.15	1	0.5	1.5	0.78	2.1
1250	0.1	1.5	0.4	1.9	0.6	1.3
1500	0.1	1.5	0.3	2.1	0.6	2.7
1750	0.05	2	0.3	2.6	0.5	2.9
2000	0	2.5	0.2	2.9	0.45	3

Fig. 2 Adam optimizer vs. ReLU, sigmoid and Tan*h* activation functions

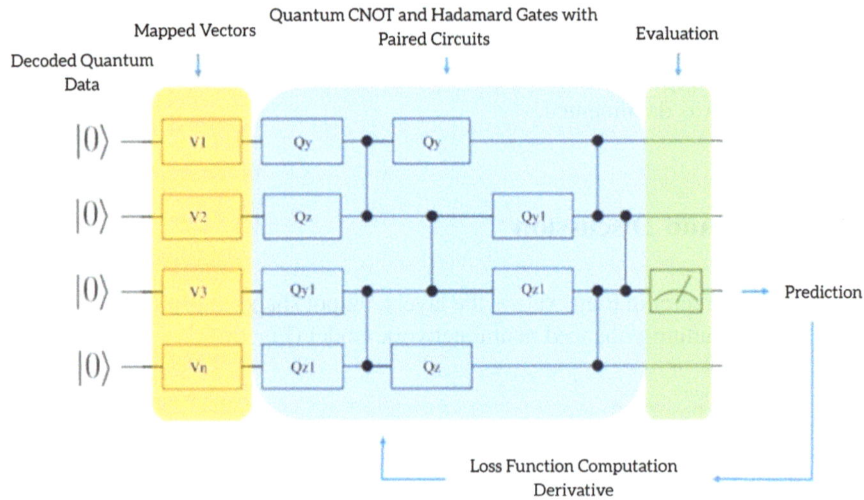

Fig. 3 Quantum neural network architecture

```
----------------------------------------------------------------
        Layer (type)               Output Shape         Param #
================================================================
            Linear-1                   [-1, 64]             128
              ReLU-2                   [-1, 64]               0
            Linear-3                   [-1, 32]           2,080
              ReLU-4                   [-1, 32]               0
            Linear-5                    [-1, 1]              33
================================================================
Total params: 2,241
Trainable params: 2,241
Non-trainable params: 0
----------------------------------------------------------------
Input size (MB): 0.00
Forward/backward pass size (MB): 0.00
Params size (MB): 0.01
Estimated Total Size (MB): 0.01
----------------------------------------------------------------
```

Fig. 4 Model summary

3.3 Training and Validation Loss

The training loss was significantly worse in the ANN model than the validation loss as compared to the QNN model. This indicated that the training dataset is more difficult to predict than the validation dataset. This was explained by the well accepted fact that the validation data percentage is lower than that of the training set.

The QNN model also performed incredibly better than the ANN model as shown by the smoother curve presented in the analysis. The ANN model data frames loading caused higher spikes at the beginning of the training curve. The spikes at the beginning of the curve were evident as the model read data at the start of the learning process. The spikes eventually flattened down after a few epochs, and the model carried on smoothly generalizing.

3.4 Training and Validation Loss Over Time

The analysis of the results showed that the QNN model performed very well as compared to the ANN one in terms of time. The QNN model average time taken for the experiment was 1.9 s as compared to the ANN time of 83.3 s. This represents a remarkable improvement in efficiency, with the QNN model exhibiting a reduction of approximately 97.7% in computational time (Figs. 5 and 6).

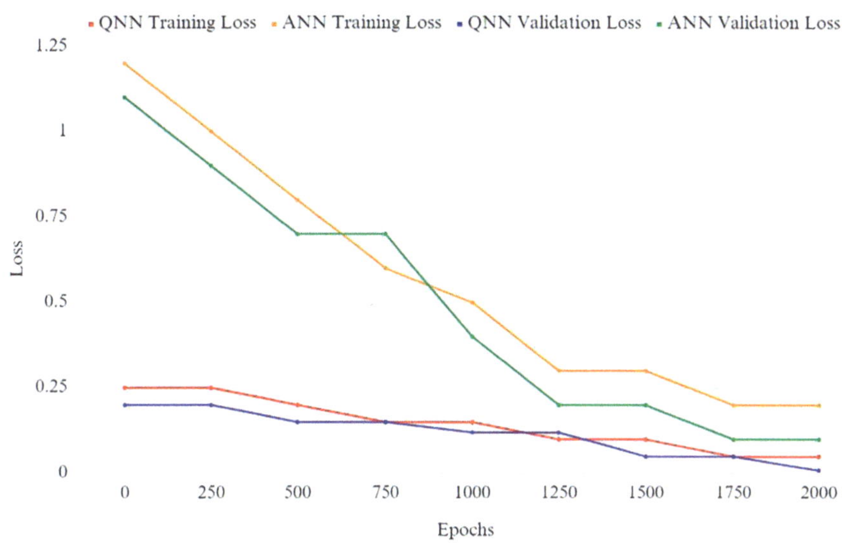

Fig. 5 Training and validation loss

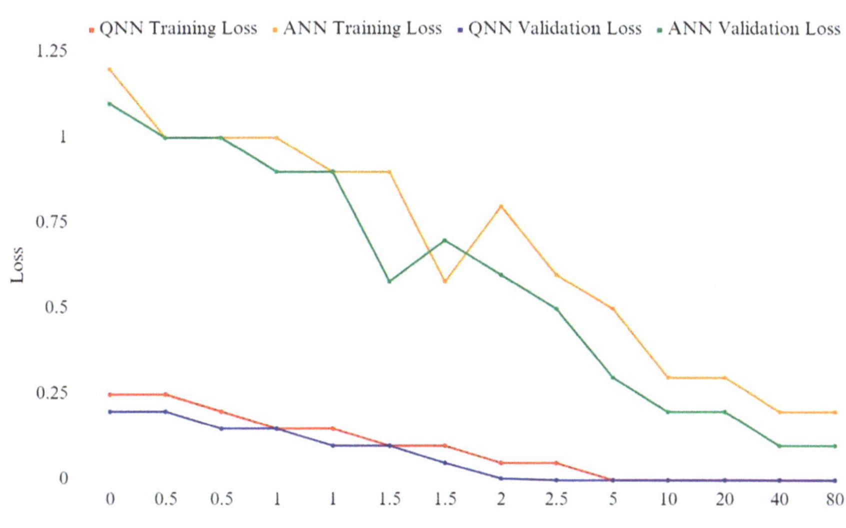

Fig. 6 Training and validation loss over time

3.5 Model Evaluation

The model was evaluated against Time, Mean Absolute Error (MAE), Mean Squared Error (MSE), Root Mean Squared Error (RMSE), Mean Absolute Percentage Error (MAPE), *R*-squared, and forecast bias (Table 2). The evaluation metrics were presented in Tables 3 and 4.

Table 2 Quantum neural network research parameters

Quantum artificial neural network algorithm								
Iteration	Training time (s)	Validation time (s)	MAE	MSE	RMSE	MAPE	R^2	Forecast bias
1	2.04	0.43	0.012	0.0003	0.02	3.18	0.997	−0.011
2	1.83	0.35	0.009	0.0003	0.02	2.18	0.998	−0.018
3	1.78	0.36	0.010	0.0003	0.02	3.78	0.998	−0.010
4	1.90	0.39	0.012	0.0003	0.01	4.45	0.997	−0.011
5	1.92	0.40	0.009	0.0002	0.01	1.79	0.998	−0.008
6	1.97	0.40	0.011	0.0003	0.02	2.18	0.997	−0.011
7	1.99	0.41	0.011	0.0001	0.01	2.39	0.998	−0.008
8	2.06	0.41	0.008	0.0002	0.02	8.97	0.998	−0.008
9	1.83	0.38	0.009	0.0002	0.02	4.81	0.997	−0.009
10	1.92	0.39	0.013	0.0003	0.02	2.24	0.996	−0.013
	Average training time	Average validation time	Average MAE	Average MSE	Average RMSE	Average MAPE	Average R^2	Average forecast bias
	1.93	0.39	0.011	0.0003	0.02	3.60	0.998	−0.008

Table 3 Neural network research parameters

Artificial neural network algorithm								
Iteration	Training time (s)	Validation time (s)	MAE	MSE	RMSE	MAPE	R^2	Forecast bias
1	83.13	6.06	30.20	29.29	35.96	4.97	0.29	−17.01
2	81.38	6.07	29.83	12.12	34.82	5.64	0.21	−14.45
3	81.77	6.06	29.59	11.65	34.14	6.10	0.16	−12.72
4	82.36	6.07	29.94	12.35	35.15	5.44	0.23	−15.22
5	87.16	6.07	29.52	11.53	33.96	6.23	0.15	−12.24
6	84.41	6.07	30.00	12.47	35.32	5.33	0.24	−15.62
7	82.52	6.06	29.99	12.47	35.31	5.34	0.24	−15.60
8	82.95	6.06	29.91	12.29	35.05	5.50	0.22	−14.99
9	84.04	6.06	30.04	12.58	35.46	5.25	0.25	−15.94
10	84.04	6.06	30.04	12.58	35.46	5.25	0.25	−15.94
	Average training time	Average validation time	Average MAE	Average MSE	Average RMSE	Average MAPE	Average R^2	Average forecast bias
	83.38	6.06	29.91	13.93	35.06	5.50	0.23	−14.97

Table 4 Forecasted QNN and ANN GDP

Year	GDP	QNN predicted GDP	ANN predicted GDP
1961	0.7929595	0.760281861	0.681540528
1973	2.5089985	2.461332083	2.065501543
1992	8.2091292	8.451481819	6.765370828
1999	12.896014	12.87607354	11.07848642
2008	35.895153	35.87516743	20.92065333
2016	74.815121	75.727211	59.84062131
2020	100.66654	101.110733	95.6920427

The QNN model exhibited substantially lower MAE values than the ANN model, as can be seen by comparing the MAE values. This indicates that the QNN model had less errors between the predicted and actual values and performed more accurately. The QML model's MSE findings had significantly smaller values, ranging from 0.0001 to 0.0003, indicating a smaller average squared difference between the predicted and actual values, as it was observed by comparing the MSE results. The ANN model, however, contained larger values with a range of 11.6537–29.2926, showing a greater average squared difference. The QNN model's MSE average value was regarded as superior because it showed a reduced error and a closer match between the predicted and actual values. A better fit of the model to the data was shown by lower MSE values, which suggested higher accuracy and precision in the predictions.

The magnitudes of the values in the QNN model were noticeably smaller than those in the ANN model. The decreased RMSE values in QNN suggested greater performance of the target variable or the forecasted values in the set. The study displayed a smaller percentage gap between the projected and actual values, with a lower MAPE score suggesting more accuracy. In the study, the QNN MAPE findings showed superiority to the ANN model's MAPE results. The QNN model has much larger R-squared values than the ANN model results. The QNN result values fall between 0.9969 and 0.9988, demonstrating a very strong fit between the regression model and the data. The R-squared values of the ANN model, on the other hand, are negative and vary from 0.2882 to 0.1491, which indicates that the model does not adequately match the data.

The QNN average forecast bias is negative (−0.00814), indicating that forecasts are generally underestimated as shown in Fig. 4. In other words, the anticipated values are slightly lower than the actual ones. The ANN average forecast bias is also negative (−14.9745), which indicates that the projections are generally underestimated. The magnitude of the forecast bias in the ANN model is greater than in the QNN model, indicating a greater difference between the projected and actual values. The QNN model appears to have a lesser forecast bias than the ANN model based only on the average prediction bias. As a result, the QNN model can be thought of as having less forecast bias because it deviates less from the actual values. Table 4 shows the forecasted GDP.

4 Conclusion and Recommendation

Based on the evaluation metrics, the QNN model demonstrates overall good performance in all areas of the forecast study. The low values of all the evaluation metrics indicate a small average deviation from the actual values. This suggests the QNN model is superior to the ANN model. This study highlights how quantum computing could revolutionize a number of industries that need intensive data processing and analysis. Therefore, the study recommends the adoption of quantum-enhanced neural networks in machine learning since it performs better in terms of MAE, MSE, RMSE, MAPE, R-squared, and forecast bias.

References

1. Ajayi, A.A., Sofowora, G.G., Ladipo, G.O.: Explaining heart failure hyper mortality in sub Saharan Africa: global genomic and environmental contribution review. J Natl Med Assoc. **112**(2), 141–157 (2020)
2. Alexeev, Y., Bacon, D., Brown, K.R., Calderbank, R., Carr, L.D., Chong, F.T., Thompson, J.: Quantum computer systems for scientific discovery. PRX Quantum. **2**(1), 017001 (2021)
3. Alvarez-Rodriguez, U., Sanz, M., Lamata, L., Solano, E.: Quantum artificial life in an IBM quantum computer. Sci Rep. **8**(1), 14793 (2018)
4. Bourassa, J.E., Alexander, R.N., Vasmer, M., Patil, A., Tzitrin, I., Matsuura, T., et al.: Blueprint for a scalable photonic fault-tolerant quantum computer. Quantum. **5**, 392 (2021)
5. Castelvecchi, D.: IBM releases first-ever 1,000-qubit quantum chip. Nature. **624**(7991), 238 (2023) Contribution review. Journal of the National Medical Association, 112(2), 141157
6. Del Vecchio, P., Mele, G., Ndou, V., Secundo, G.: Creating value from Social Big Data: Implications for Smart Tourism Destinations. Inf Proces Manag. **54**(5), 847–860 (2018). https://doi.org/10.1016/j.ipm.2017.10.006
7. Djomo Nana, J.M., Epo, B.N.: Road fatalities and extreme poverty in SubSaharan Africa: how fatal is it for the breadwinners? J Poverty. **122** (2023)
8. Doan, H.Q.: Trade, institutional quality and income: Empirical evidence for sub-Saharan Africa. Economies. **7**(2), 48 (2019)
9. Ertel, W.: Introduction to artificial intelligence. Springer (2018)
10. Ghani, N.A., Hamid, S., Hashem, I.A.T., Ahmed, E.: Social media big data analytics: a survey. Comput Human Behav. **101**, 417–428 (2019)
11. Hardy, K., Maurushat, A.: Opening up government data for Big Data analysis and public benefit. Comput Law Secur Rev. **33**(1), 30–37 (2017)
12. Kouton, J.: Relationship between economic freedom and inclusive growth: a dynamic panel analysis for sub-Saharan African countries. J Soc Econ Dev. **21**(1), 143–165 (2019)
13. Mahomed, Z.: PPPs or iPPPPs: which model provides greater infrastructural social impact in sub-Saharan Africa? IF HUB, Issue 1, March, 3–7. Retrieved from ResearchGate (2022, April)
14. Mangini, S., Tacchino, F., Gerace, D., Bajoni, D., Macchiavello, C.: Quantum computing models for artificial neural networks. Europhys Lett. **134**(1), 10002 (2021)
15. Morrison, L.T., Anderson, B., Brower, A., Talbird, S.E., Buell, N., MacDonald, P.D., Honeycutt, A.A.: Macroeconomic impact of Ebola outbreaks in Sub-Saharan Africa and potential mitigation of GDP loss with prophylactic Ebola vaccination programs. PLos One. **18**(4), e0283721 (2023)

16. Njindan Iyke, B., Odhiambo, N.M.: Foreign direct investment and trade openness in Sub-Saharan Africa: Evidence from non-linear panel cointegration. Res Int Bus Finance. **51**, 101074 (2020)
17. Qiu, J., Wu, Q., Ding, G., Xu, Y., Feng, S.: A survey of machine learning for big data processing. EURASIP J Adv Signal Process. **2016**, 1–16 (2016)
18. Qolomany, B., Al-Fuqaha, A., Gupta, A., Benhaddou, D., Alwajidi, S., Qadir, J., Fong, A.C.: Leveraging machine learning and big data for smart quantum-communications. Accessed, 6(12), 2021. Springer (2019)
19. Sze, V., Chen, Y. H., Emer, J., Suleiman, A., Zhang, Z.: Hardware for machine learning: Challenges and opportunities. In 2017 IEEE Custom Integrated Circuits Conference (CICC) (pp. 1-8). IEEE (2017, April).
20. Talia, D., Trunfio, P., Marozzo, F.: Research trends in big data analysis. In: Data Analysis in the Cloud, pp. 123–138 (2016)
21. Ukhalkar, P.: The transformative potential benefits of big data in government and public sector domains. Int J Adv Sci Res. **3**(1), 30–33 (2018)
22. Wang, G., Liu, X.: Predicting China's economic growth using machine learning and big data. Appl Econ. **52**(28), 3053–3066 (2020). https://doi.org/10.1080/00036846.2020.1738687
23. Wani, M.A., Bhat, F.A., Afzal, S., Khan, A.I.: Advances in deep learning. Springer (2020)
24. Zhou, L., Pan, S., Wang, J., Vasilakos, A.V.: Machine learning on big data: Opportunities and challenges. Neurocomputing. **237**, 350–361 (2017)
25. Zhou, X., Qiu, D., Lou, L.: Distributed exact quantum algorithms for Bernstein-Vazirani and search problems. arXiv preprint, arXiv:2303.10670 (2023)

Index

A
AI algorithms, 88, 93, 99, 100, 102, 108, 152, 153, 156, 168, 169, 176, 178, 229–232, 234, 240–244, 271, 272
AI sensors, 229–244
Artificial intelligence (AI), 2, 34, 85, 149, 161, 185, 226, 229, 268, 294
Artificial neural networks (ANNs), 114, 119, 120, 128, 130–133, 135, 185, 294, 298–302

B
Biomass wastes, 113–135
Blockchain, 236, 238, 244

C
Calibration, 216–217, 225, 239, 242, 243, 254, 263
Circular economy (CE), 113, 117, 119, 122, 143, 144, 147–148, 152–153, 155, 156
Citizen science, 22, 25, 27
Classical approaches, 185, 186
Co-gasification, 113–135
Computer vision, 2–24, 26, 27, 69–72, 74
Control valves, 90, 91, 161–183
Convolutional neural networks (CNNs), 3, 33, 70, 150, 173, 230, 276

D
Data augmentations, 15, 22, 23, 26, 49, 268
Data protection, 206
Deep learning (DL), 3–5, 8, 11, 20, 22, 26, 33–38, 40, 42, 48, 62, 63, 70, 71, 74, 75, 81, 173–174, 181, 187, 190, 200, 201, 230, 268, 271, 277, 281, 283, 294
DHT22 sensor, 254, 257, 258, 263
Domain adaptation, 18

E
Economic forecasts, 295
Edge computing, 17, 22, 24, 25, 98, 99, 106, 179–181, 229, 233, 244
Energy efficiency, 98, 204, 209, 230–235, 238, 242–244
Energy harvesting, 232, 234, 243–244
Environment, 5, 34, 86, 128, 143, 163, 204, 229, 247, 294
ESP32, 206–210, 214, 215, 247–265
Ethical AI, 24, 25

F
F1, 16, 48–52, 56, 75, 77–81, 186, 189, 194, 195, 197–199
Face masks, 69, 70, 74, 81
Face recognition, 69–81

P. Naidoo, M. Sibanda (eds.), *Measurement, Automation, and Control in Artificial Intelligence and Machine Learning*, Signals and Communication Technology, https://doi.org/10.1007/978-3-031-91802-5

G
Gaussian process regression (GPR), 114, 119, 120, 128, 130–132, 134, 135
Gradient boosting regression (GBR), 114, 119, 128, 130–135

H
Home Energy Management Systems (HEMS), 205
Hypothesis testing, 220, 221

I
Image processing, 3, 36
Inception, 11–12, 26, 72, 75, 103
Industry 4.0, 94, 97–98, 113–135, 148
Internet of Things (IoT), 121, 135, 144–148, 203, 215, 231, 243, 248, 251

L
Long-range (LoRa), 38, 145, 155

M
Machine learning (ML), 2–27, 33, 38, 44, 69–71, 73, 74, 85–110, 113–135, 149–151, 154, 161–183, 188, 200, 226, 235, 237–238, 240–243, 268, 282, 284, 294, 295, 302
MATLAB, 128, 223
Model interpretability, 38
Multi-modal data integration, 18, 21, 25, 27

N
Neural Machine Translation (NMT), 185, 186, 188, 200
Neural networks, 11, 12, 33–64, 96, 97, 154, 171, 185, 278, 293–302

P
Precision, 2, 34, 75, 109, 143, 161, 186, 210, 232, 247, 301
Predictive modeling, 113–135
Process automation and control, 85–110, 162, 168, 183
Pruning, 23, 26, 35, 36, 38–40, 42–49, 51, 62, 64

Q
Quantization, 23, 26, 35, 38, 40–42, 45–49, 51, 62, 64
Quantum computing, 243, 302
Quantum-enhanced neural network, 296, 297
Quantum machine learning, 150

R
Raspberry Pi, 251, 252
Real-time processing, 22, 24–27
Recall, 16, 48–52, 56, 75, 77–81, 186, 189, 194, 195, 197–199
Residual networks (ResNet), 10, 11, 26, 45, 49, 51, 52, 55, 75, 279

S
Saliency maps, 268, 276, 277
Scalability, 5–7, 23–26, 100, 128, 232, 234, 236, 271
Sensor calibration, 217, 239
Smart metering, 203–226
Support vector machine (SVM), 69–81, 119, 120, 134, 150, 171
Sustainability, 2, 98, 100, 115, 121, 122, 143–156, 231, 232, 242
Syngas compositions, 114–116, 119, 120, 124, 127, 128, 130–132, 135
Synthetic data, 107, 268, 278–279, 284–287

T
TBNet, 37, 38, 62, 63
ThingSpeak, 252, 253
3D printing, 212, 258
Transfer learning (TL), 18, 22, 23, 72, 75, 174, 196, 200, 268, 278–280, 284–287
Tuberculosis detection, 33–64

W
Wildlife detection, 2–27

X
X-ray image classification, 34, 57, 60

Y
You Only Look Once (YOLO), 8–10, 13, 18, 19, 26

MIX
Papier aus verantwortungsvollen Quellen
Paper from responsible sources
FSC® C105338

If you have any concerns about our products,
you can contact us on
ProductSafety@springernature.com

In case Publisher is established outside the EU,
the EU authorized representative is:
Springer Nature Customer Service Center GmbH
Europaplatz 3, 69115 Heidelberg, Germany

Printed by Libri Plureos GmbH
in Hamburg, Germany